THE CASH LANDRUM INCIDENT

JOHN F. SCHUESSLER

The Cash Landrum Incident
Copyright © 1998/2022- by John F. Schuessler

All rights reserved. No part of this book may be reproduced in any form without written permission from the publishers unless by reviewers who wish to quote brief passages.

Manufactured in the United States of America

ISBN 978-1-953321-11-4

Foreword by Bob Pratt
Book design by Jane McWhorter
Cover design by Deborah Herman

Published by

UFO Books Press
A Division of Micro Publishing Media, Inc
PO Box 1522
Stockbridge, MA 01262

DEDICATION

To Betty Cash, Vickie Landrum, and Colby Landrum for their suffering and perseverance. This is the story of their ordeal.

Contents

Acknowledgments 6
Foreword 8
Preface 10
Prologue 13
Chapter 1   The Encounter 15
Chapter 2   The Fire Goes Out 22
Chapter 3   The Trip Home 24
Chapter 4   The Sickness Begins 28
Chapter 5   Betty Is Hospitalized 34
Chapter 6   Vickie Toughs It Out 39
Chapter 7   The Investigation Begins 44
Chapter 8   The First Documented Report 48
Chapter 9   Betty Cash Is Interviewed 53
Chapter 10 Vickie Tells Her Story 58
Chapter 11 Colby Answers Questions 63
Chapter 12 Sensory Data—Sounds & Heat 67
Chapter 13 Helicopters Are Up There Too 75
Chapter 14 Related Ufo Activity 78
Chapter 15 Baseline Health Status 84
Chapter 16 Betty's Medical Condition Is Defined 92
Chapter 17 Betty Is Hospitalized Repeatedly 98
Chapter 18 Dr. Shenoy Speaks Out 104

Chapter 19	Another Doctor Offers To Help	107
Chapter 20	Hot Air Balloon Not The Culprit	113
Chapter 21	Lack Of Radar Hampered The Investigation	115
Chapter 22	Troubled Times	120
Chapter 23	Hypnosis Was Used	127
Chapter 24	The Pentagon Expresses An Interest	141
Chapter 25	Ellington Commander Responds	145
Chapter 26	A Helicopter Comes To Dayton	148
Chapter 27	The Helicopter Pilot	154
Chapter 28	The Army's Investigation	157
Chapter 29	Police Officer Saw Helicopters	162
Chapter 30	Other Reports Of Helicopters	168
Chapter 31	Unconfirmed But Interesting Reports Of Helicopters	175
Chapter 32	The Legal Battle Begins	178
Chapter 33	Federal court action begins	182
Chapter 34	The Process Of Discovery	185
Chapter 35	The Interrogatories Are Poorly Answered	192
Chapter 36	The Docket Call	197
Chapter 37	Wrapping Up The Case	203
Chapter 38	The Pain Continues	209
Addendum		219
About the author		273

Acknowledgments

This book would not have been possible without the continuous encouragement and support of my wife, Kathy. Throughout the investigation, she transcribed audio tape recordings, reviewed the investigator's notes and made suggestions for improvements in the ongoing investigations, developed paintings depicting the UFO event, critiqued articles and reports, and reviewed the manuscript of this book. She was patient and understanding about the hundreds of hours I spent away from home investigating the event and assisting the victims. This included foregoing our annual vacations because I had expended my annual leave allotment on this case.

I am grateful to Bob Pratt for his extensive participation in this investigation. Bob conducted interviews, searched for additional witnesses, worked with me on the original book manuscript, which was never published, and always tried to be helpful to the victims of the event. I believe Bob is one of the most thorough and caring investigators in this field.

Expert help during the initial investigation of this case was provided by Bill Eatwell, Alan Holt, Dave Kissinger, Don Tucker, and other members of the Houston-based Vehicle Internal Systems Investigative Team (VISIT). Their work helped to establish a firm baseline for the years of investigation that were to follow.

Many thanks go to" my friend and confidant, Dr. Howard Sussman. His ability to assimilate large quantities of information and then suggest strong avenues of inquiry is second to none.

Medical professional guidance was provided by Dr. Peter Rank, Chief of the Department of Radiology at Methodist Hospital in Madison, Wisconsin. During his life, he helped a lot of people, but his caring support and professional suggestions for helping Betty and Vickie were wonderfully helpful. They really made a difference to Betty and Vickie.

A special note of gratitude goes to Walter H. Andrus, Jr., International Director for the Mutual UFO Network, Inc., for his consistent support and friendship throughout the years.

Finally, I want to thank Betty Cash, Vickie Landrum, and Colby Landrum for sharing their experiences so that others may not need to go through the same agony they experienced. This book was written at their request to be used as a guide to help the next victims when this type of event occurs again.

Foreword

Since the mid-1980s, many American UFO researchers have been mired in endless quarreling over the Roswell incident, MJ-12, and other quagmires, with little to show for it other than bitterness. In doing so, they've overlooked or ignored one of the most significant cases ever to come to light.

If all the energy and time spent on those ugly battles had been devoted to finding out who or what was behind the Cash-Landrum incident, we might now know what the U.S. Government really knows about UFOs and perhaps be closer to solving the UFO mystery. We might also have found better ways to treat the health problems of the three victims in the case—Betty Cash, Vickie Landrum, and Vickie's grandson, Colby Landrum—and alleviate their suffering.

Since 1975, I have been fortunate enough to study UFO reports in the United States and ten other countries, speaking to at least 1,800 people who had sightings or encounters. Since 1981, I have concentrated mainly on one country, Brazil, because so many harmful things happened to people in UFO encounters there, and that aspect of the phenomenon particularly interested me. Nearly every one of the several hundred Brazilians I have interviewed feared for their lives during their UFO experience. Some were even injured, several permanently, and a few died.

However, as intriguing as all of those cases were, none of them quite matched what happened to Betty, Vickie, and Colby when they encountered an unidentified flying object on a lonely Texas highway one night in December 1980. They, too, felt terrorized during the incident and feared the world was coming to an end. This case was unique, one of extremely high strangeness with multiple witnesses, serious medical effects, military involvement, and government cover-up.

On that dark night in 1980, a large, strange machine swooped down from the sky and then hung noisily just above the highway in front of Betty, Vickie, and Colby, blocking their way for fifteen minutes. By the time it left, it had exposed them to emissions of some kind that seriously damaged the health of all three.

Their suffering was compounded by the fact that when the UFO went away, it was pursued by nearly two dozen military helicopters—meaning the government was involved with the mysterious object and knew what it was and where it went. Yet, even though other people saw or heard the choppers, the government denies any of its helicopters were in the area that night.

I am very familiar with the Cash-Landrum incident. I had many lengthy interviews with Betty and Vickie and found them to be very credible people, as have all other investigators who have looked into the case. Even an Army colonel who investigated their claims for the Pentagon found them to be credible.

I also talked to other people who helped authenticate their claims, and I saw both women not long after the encounter when they had lost much of their hair and were still trying to recover from the skin and other health problems caused by the encounter. I do not doubt that what they say happened did happen.

We are fortunate that John Schuessler headed the investigation of this case. He is one of the most capable, serious, resourceful, and open-minded researchers in American ufology, a patient and objective man who is not prone to jumping to conclusions or embracing lunatic theories. Several years after the incident occurred, John shared virtually everything he learned about this case. I'm still amazed at the extraordinary amount of pertinent data accumulated. The result is that this is one of the most thoroughly investigated cases in ufology.

This case is significant because of the enormous size and shape of the craft and its strange behavior over the highway, the serious detrimental effects its presence had on the three main witnesses, the fact that the government was very much involved in this incident, and that those who knew about it chose to cover it up.

The Cash-Landrum incident has to be one of the most thoroughly investigated cases in the history of ufology. I, myself, spent hundreds of hours interviewing Betty, Vickie, and other witnesses and studying the case, but my efforts pale when compared to what John Schuessler has done. He has devoted years to investigating every aspect of the case—

Preface
The Legacy of the Cash-Landrum UFO Incident

By Barbra Sobhani

The Cash-Landrum UFO Incident, self-published by my father, John Schuessler in 1998, is a comprehensive look at one of the most intriguing UFO cases in the United States in recent times. On December 29th, 1980 in East Texas, Betty Cash, Vickie Landrum and Colby Landrum experienced something that would change their lives and destroy their health. John Schuessler was the chief investigator of this case, but it goes beyond that. My Dad was personally invested in helping these people and as his daughter, I experienced how the case moved from a simple investigation of a UFO sighting, to a lifelong pursuit of truth.

All three of the witnesses fell ill immediately after the sighting. They wanted to know what was wrong with them and went to the military seeking help. Finding no help there, they turned to MUFON, contacting my father who was working at the NASA Johnson Space Center, hoping to find some help and some answers. His training as an engineer and scientist, his connections to medical professionals, and his many years of investigating cases all played an important role in dealing with this case.

One of the unusual things about the case is the sighting of the military helicopters surrounding the craft, which they appeared to be chasing or guiding. This was a concrete lead to follow, double rotor helicopters (CH-47) were not commonly used in civilian operations at the time, so the sighting of these gave Betty and Vickie a place to start when they realized they were going to need help. This key point indicated military involvement, or at least knowledge of the incident. If the proximity to the UFO caused such harm to Betty, Vickie and Colby, it was logical to assume the pilots might have experienced similar effects. The official military response was that helicopters did not fly on Mondays, so they could not have been present that night. This ridiculous statement was disproven time and again as my father documented and photographed CH-47s in the sky on many a Monday following the incident. I have distinct memories of watching the skies and attending air shows trying to track down helicopter pilots. At one particular air show, a pilot identified as being involved was avoiding being interviewed by my dad, but I was able to get some photos since I was just a kid.

After years of trying to get help from the government, due to their apparent involvement in the incident, there was never a resolution to their case. However, since that time, we have witnessed the recent disclosures of Navy pilot UFO encounters. There have also been other documented, similar injuries associated with UFO encounters from workers at Skinwalker Ranch. Other unusual effects from a more sophisticated energy form (referred to as 'Havana Syndrome') at the US embassies starting in Cuba, having effects similar to those that ruined the health of Betty, Vickie and Colby.

Betty Cash battled health issues for the rest of her life. My Dad remained close to her, corresponding regularly. Betty passed away on December 29, 1998, due to complications from anemia, a stroke and multiple surgeries. Vickie Landrum also struggled with her health until her death on September 12, 2007 at the age of 83. Colby Landrum has had similar health issues throughout his life. He currently lives in Texas and generally stays out of the public eye.

This book contains many of the original photographs, interview transcripts and documents involved in the case investigation. This case endures as a classic UFO case because of the amount of evidence available, particularly the medical documentation. It is not a case easily dismissed as a "light in the sky".

John Schuesssler is one of the founders of MUFON, which celebrated its 50th anniversary in 2019. He also founded and ran Project VISIT, specializing in the technological investigation of events. He has pursued the truth behind these events in hopes of helping these individuals medically. If we could better understand the technology and energy source of the craft, perhaps better treatment could be devised. His interest in medical evidence connected to technology in sightings persisted throughout the years. My Dad is considered one of the finest UFO investigators and he continually advocates for scientific research of the subject. The Cash-Landrum incident was a dominant feature in my life as I grew up, and I feel it is important to have this information available.

- the encounter itself,
- tracking down other witnesses,
- checking the medical records and histories of the three witnesses,
- consulting medical authorities,
- helping Betty, Vickie, and Colby get medical and legal help,
- probing the military and government angle and, most importantly,
- keeping the case alive when others had abandoned it.

The case remains one of the most important in the history of ufology. It has never been resolved. Someone in the U.S. knows exactly what happened to Betty and Vickie, and Colby, but they have remained silent despite all the suffering—and ridicule—those three endured.

Hopefully, this book will keep the search for truth alive, and one day we'll know not only what happened that night but also why those in the know chose to sacrifice the health and reputations of Betty, Vickie, and Colby.

Bob Pratt
Lake Worth, Florida
Robert V. Pratt (1926-2005)

Bob Pratt began as a skeptic writing for such tabloids as the National Enquirer until May of 1975 when he interviewed over 60 people who convinced him that UFOs are real. Bob Pratt was a newspaper and magazine reporter and editor for more than four decades. Bob began investigating UFO sightings in 1975, traveling to many countries, including Japan, Canada, Mexico, Puerto Rico, Argentina, Bolivia, Peru, Chile, Brazil, and Uruguay. Bob interviewed many witnesses and gathered data on sightings, writing many articles throughout the year. Bob served as the editor for the *MUFON UFO Journal*. He is the author of *UFO Danger Zone*, documenting his numerous trips to South America to investigate UFO Sightings.

Prologue
The Intruder Arrives—a Likely Scenario

The helicopter amphibious assault ship U.S.S. New Orleans is waiting in the early evening darkness for the return of the *Operation Snowbird* helicopter task force. Moving at a speed of less than 5 knots, the New Orleans is heading for the rendezvous point in the Gulf of Mexico, several miles south of Crystal Beach, Texas. They had been monitoring the activities of an intruder for more than 24 hours, and the waiting would soon come to an end.

Meanwhile, the helicopter task force had flown inland, crossing the Louisiana coastline a few miles east of Port Arthur, Texas. Turning west, they fly between Port Arthur and Beaumont, carefully avoiding any large towns.

After their failure to rescue the American hostages in Iran in April, the task force has been completely restructured while planning for a second rescue attempt. The new unit of helicopters included Black Hawks, Chinooks, a Sky Crane, and several light observation models. All but the light observation craft had been modified at an Army depot in Pennsylvania by adding special communications, infrared radar guidance, night vision systems, refueling systems, and special long-range fuel tanks. In November, they had been assigned to a unique training exercise that included night precision flying in Texas in conjunction with units from Grey Field at Fort Hood. It was just dumb luck that the New Orleans had been in the Gulf of Mexico when the intruder arrived. Because of their plan to try again to rescue the hostages in Iran, training had not been suspended for the holidays.

The U.S. Government was keenly aware of the intruder's operations over the continental United States. Ever since the Holloman Air Force Base landing in New Mexico in 1964, government officials had been aware of all major alien intrusions; but only interfered when there was a threat of exposure.

Just 24-hours earlier, civilians reported the flight of the large diamond-shaped craft in the skies over Arkansas, and they said it was heading south. As usual, however, government officials just made fun of the reports. Now it was the night of December 29, 1980, and the intruder had concluded its operations over Louisiana and was heading west in Texas. As the huge craft approached the town of Liberty around 8 p.m., it sustained a major systems malfunction. Unless repairs could be made quickly, it would go down somewhere in the East Texas Piney Woods.

The intruder's emergency signal was picked up and relayed to all operating units in Texas and Louisiana via the NORAD network in Cheyenne Mountain, Colorado. In minutes, helicopter units from New Orleans and from Fort Hood were airborne. If the intruder did crash, they were prepared to cordon off the area to keep civilians out while the clean-up operations were underway. The Chinooks carried troops and chemical/biological warfare equipment. The Skycrane was ready for heavy-lift removal operations, while the Blackhawks would control the airspace over and around the crash site.

As the intruder passed near Dayton, Texas, its systems were still malfunctioning, and a crash appeared imminent. Ground-zero was pinpointed as a site about three miles from Huffman, Texas, a small town to the northeast of Houston. It would be nearly 9 p.m. before the New Orleans helicopter task force could reach the crash site and several minutes later for the Fort Hood unit. The plan was to either escort or haul the intruder to a safe zone over the Gulf of Mexico. But from miles away, they could see the bright glow from the malfunctioning intruder and were prepared to initiate clean-up operations when they arrived at Huffman. They were not aware that someone on the ground was already in harm's way.

NOTE: The scenario is a tool used by Futurists and military planners to determine possible outcomes in world events. Based on all available facts, this scenario was chosen over eight other scenarios as the most likely description and explanation of what happened the night of December 29, 1980, when Betty Cash, Vickie Landrum, and Colby Landrum sustained life-threatening injuries during a close encounter with a huge diamond-shaped object in east Texas.

CHAPTER 1
The Encounter

The glow appeared on the horizon so gradually that several minutes passed before anyone in the car noticed it. It was a vertical streak of red, and even though it appeared to be miles away, it stood out clearly in the dark sky.

Little Colby Landrum was the first to spot it.

"What's that, Grandma?" he said. He was standing behind the front seat of the 1980 Oldsmobile Cutlass behind Betty Cash, 52, who was driving, and his grandmother, Vickie Landrum, 57. Colby, a seven-year-old with vivid blue eyes and hair so blond it was nearly white, nodded toward the streak in the sky as he spoke. But neither woman noticed his gesture. He had his arms crossed on the top of the front seat and was quietly listening to the running conversation

Sighting location along FM 1485, this is the straight section of road where the first sighting occurred.

between the two women.

Both had heard Colby, but they thought he was asking about something one of them said because they had been telling him what life was like before he was born, or what Vickie called "used-to-be times."

Colby continued to listen, but his curiosity about the thing in the sky was growing. It appeared to be getting larger.

It was shortly before 9 p.m. on December 29, 1980, a Monday, and they were eight or nine miles east of New Caney, Texas, driving south on State Farm Road 1485 en route to their hometown of Dayton, about 15 miles away.

Dayton is a small town, one of many that support the area's ranching and oil workers. It is a half-hour's drive northeast of Houston.

They had gone out earlier in the evening to find a bingo game, driving 23 miles north to the town of Cleveland, where the VFW club nearly always has a game on Monday night. But the club was dark and drove back south to New Caney, where the American Legion hall also holds games. Again, they were disappointed. The Legion Hall was closed. They gave up trying to find a game and drove to a truck stop on the east side of New Caney for something to eat. They were there nearly an hour and left about 8:45 p.m., heading east toward Dayton on State Farm Road FM1485. The highway was dark and deserted.

Winters in southeastern Texas are generally mild, with the temperature dipping down as low as the mid-20s only three or four times a year. On this night, the rain and drizzle that had been falling during the day had stopped, and the air was a chilly 40 degrees.

Inside the car, however, it was warm and comfortable. They had the heater on, and the radio tuned to the country-western sounds of Radio KIKK. The radio was playing softly and was virtually ignored as the women talked.

They were driving through a section of east Texas known as the Piney Woods, which remains green and fresh while the dense undergrowth of weeds and vines turns a crisp-brown in the winter.

The highway was hemmed in by walls of tall pine trees, with only a small clearing here and there for a solitary house or trailer home. During the daytime, the road is heavily traveled by logging trucks, but Betty's car seemed to be the only vehicle on the highway on this night.

The red glow on the horizon was getting bigger now, and Colby became more curious. Betty was saying something to Vickie, with her head half-turned toward her, when Colby reached over and nudged Betty's face to the left.

"Aunt Betty, what's that?" he asked.

"What, Honey?" Betty replied, glancing at him.

By now, Colby had gently pushed his grandmother's face so that she was looking at the sky ahead. "See, Grandma? What's that?"

Both women had seen the light, but they hadn't paid any attention to it. Now they stopped talking and studied the bright light. "Boy, that's strange," Betty said.

To Vickie, it looked like a long streak of fire. "Well, I know it isn't a plane or a star," she said, "but I can't figure it out."

"Maybe two planes collided," Betty said, but as soon as she spoke, she knew that couldn't be the explanation. Then she thought that perhaps a new shopping center had opened in a nearby town and that a searchlight was being used to draw a crowd. But whatever was in the sky didn't move around the way a searchlight does.

"I don't know what it is, Vickie, but it's causing the sky to light up," Betty said.

Vickie said to Colby: "Honey, I don't know what it is, but don't worry about it. It's nothing to worry about."

Betty had been driving about 50 to 55 miles an hour, but she began to drive more slowly. Whatever it was, it was definitely closer now. Whether it was coming toward them or standing still as they approached it, they couldn't tell.

They drove another two or three miles, losing sight of it briefly at times as the road curved and trees got in the way. But it was getting bigger and brighter all the time.

"Maybe it will just go on," Betty said, becoming more apprehensive. "Maybe we won't see anything else." In a matter of just several minutes, it had grown from a small red streak on the horizon to a fiery-looking thing the size of a car, and they began to become concerned. Colby was now quite frightened. Then, with no warning at all, the sky seemed to split open, and the object came angling down directly in front of them, settling swiftly between the trees just ahead of them and above the highway. It seemed enormous!

"Oh my God!" Betty shrieked, "What is that thing?"

Vickie was shaken and frightened, and Colby was terrified. The fiery object was only a few hundred feet in front of them, and they were rapidly getting closer to it. It towered above the trees, with the bottom coming more than halfway down into the trees. Although it was still well above the highway, strips of flames blasted down from the bottom, virtually blocking the way, and a roaring sound bombarded their ears. It gave off dazzling brilliance. Almost immediately, they began to feel the heat in the car. "Oh, Lord Jesus!" Vickie cried.

Colby screamed and clawed his way over into the front seat to get into the protective arms of his grandmother, burying his face in her lap.

Betty's first panicky impulse was to step on the gas and drive under the thing to get away from it, but Vickie shouted: "My God, Betty! Stop! You're going to run into it, and it's going to burn us up."

Betty slammed on the brakes bringing the car to a halt in the middle of the highway less than 150 feet from the object. Vickie had to brace herself against the dashboard to keep her and Colby from being thrown forward.

"If we had gone on, I know we would have burned up," Vickie shouted over Colby's wailing. He was frantic, crying and begging her to hold him.

Betty glanced to the side and then looked back through the rear window, seeking some way to escape, but the highway was narrow, and she was afraid of getting stuck in the muddy ditches if she tried to turn around.

"Vickie, I can't even see the sides of the road!" she shouted. "I can't turn around, and I don't dare back up." She had never been so frightened in her life. She didn't know what to do. "Oh, Vickie, what's going to happen to us?" she cried.

"I don't know," Vickie cried in despair. "It's like the whole woods is on fire."

Colby was hysterical and Vickie, trying to soothe him, kept saying: "Colby, baby, it's going to be all right, it's going to be all right." But she was terribly frightened herself and, raising her voice over his screams and the noise from the object, she said to Betty: "It looks like the end of the world!"

It was getting warmer and warmer, and Betty opened her door and got out. She was going to run away from whatever this thing was, but because of the brilliant light, it gave off., she couldn't see where to run to. She felt trapped and helpless as she stood behind the car door.

Then Vickie got out, driving Colby into new hysterics. "Grandma! Don't go!" he screamed, clawing at her, trying to pull her back in. He thought she was going to leave him, and in desperation, he screamed anew, jumped out of the car, and tried to run. But Vickie caught him and held him close to her, hemming him in between her and the front seat.

Vickie wanted to run too, but she wasn't going to leave Colby behind, and she knew she couldn't run and carry him too. The trio stood behind the doors and stared at the object. It was nothing like they had ever seen or imagined before. Vickie had her left hand on the top of the car as she stood there in awe, an action she would later regret.

The bright glare was nearly blinding. It seemed to come from the flames that periodically gushed from the bottom of the object, but they couldn't really tell what was causing it.

By shielding their eyes, they could see beyond the brilliant luminescence what appeared to be a grayish, metallic structure as big as the 200-foot-tall water tower in Dayton. It was so huge they were afraid it would fall over on top of them.

It was more or less diamond-shaped, coming to a point at the sides and tapering to a blunt end on the top. They couldn't see the bottom because of the bright light and the flames.

Flames blasted down continuously but didn't touch the highway. From time to time, the flames shot down with even greater intensity, and at these times, they could hear a loud whooshing sound, similar to the noise made by air brakes on a truck. The object would rise twenty or thirty feet whenever this happened but then would settle back down when the flames subsided. Sometimes the flames seemed to go out almost entirely, but the brilliance of the object never diminished.

The object seemed to be struggling to rise above the trees, but it would go up only so far, and then it would settle back down again. As it did this, it emitted a shrill beeping sound at irregular intervals, so loud that it hurt their ears.

Colby, still crying uncontrollably, was so distraught that he fought to break free of Vickie, but she kept a firm grip on him. After a minute or two, he got back into the car and begged Vickie to get back in with him, nearly pulling her clothes off as he tried to get her back into the car. She kept a grip on him as she continued watching the object, but after two or three more minutes, he had become so terror-stricken that Vickie became alarmed, and she got in the car with him.

"Momma's back in the car," she told him, trying to calm him down. "Momma loves you, darling." She had raised him since he was a toddler, and she was both mother and grandmother to him.

Vickie hugged him tightly. His heart was beating so wildly that she could feel it through her sweater. He was having trouble breathing, and Vickie was afraid he was going to have a heart attack. "Colby, honey, Colby," she kept saying over and over, trying to calm him down. "Baby, you look real close and right at the center of that thing. If you see a big man, come out, that's going to be Jesus. He's not going to hurt you. He'll carry us to a better place. Don't be afraid. You look straight at it and just keep your eyes on it, and when you see something come out of it, it will be Jesus. He's not going to hurt us. He's coming after us, and there's no way that God will ever hurt anybody."

She kept talking to Colby, and gradually, he pulled his face away from her chest and looked at the flaming object. After about a minute, he stopped crying enough to ask: "What about Pap-paw?" That is what he called his grandfather.

Vickie replied: "We'll go through Dayton and pick him up. We'll go by and get Poppa. Don't you worry about him. We're going to carry him with us."

That seemed to calm Colby down a little, but then Vickie realized that Betty, instead of getting back into the car with them, had actually begun to walk toward the front of the car.

"Betty," she screamed, "Get back here before you burn up."

Betty could just hear her over the roar, but she did not turn back, thinking that if she got a little closer to the object, she could get a better look at it. She moved slowly, a foot or two at a time, unsure just where she was because it was difficult to see anything. Once, she reached out with her right hand to touch the fender and instantly snatched the hand back. The metal was blistering hot.

She raised her left arm to shield her eyes, but she still couldn't make out any details on the object. By now, her eyes were feeling as if someone had pulled a thin veil over them. She wasn't sure just where she was but thought she was near the car's front bumper. She didn't dare go any further.

She stood there for a long time, at least three minutes, and perhaps as much as five. All the time, she could hear Vickie screaming at her to get back in the car.

The heat finally became so unbearable that she turned and walked back to the door. She started to grab the door handle, but it was too hot, and she had to use the tail of her leather coat to open the door.

"Why in God's name did you do that?" Vickie demanded angrily when Betty was back inside the car. "You could have burned up out there!"

"I don't know, " was all that Betty could say. "I thought maybe I could get a better look, I guess."

"Well, you shouldn't have gone out there," Vickie replied, still quite upset.

Vickie and Colby perspired freely, but Betty's skin was only mildly damp even though she still had her leather jacket on and had been outside longer. Both women had been crying for some time, and Colby had never stopped.

They didn't know how long they had been there. It had been fifteen minutes, perhaps, and the object still lumbered slowly up and down, with no change in the heat, the roaring, the beeping, or the brilliance.

"There's no way we can ever live through this heat," Betty sobbed. "We're going to die."

Vickie was too busy praying to God to answer, and Betty began to pray quietly.

Then suddenly, another mass of flames shot down from the object and in an instant, it rose swiftly above the trees, almost out of their line of vision. The object was leaving.

CHAPTER 2
The Fire Goes Out

They were surprised by the speed of the thing and immensely relieved to see it go, but they still didn't feel safe. At first, they could no longer see the object, although the woods and highway were still brightly lit. They had to lean forward and look up through the windshield to see the object. It had risen to a height of perhaps 600 to 800 feet above the ground. They could no longer distinguish its shape.

Then it stopped moving up, and the flames died out completely. Almost immediately, they became fearful again. But the object appeared to tilt over on its side and then began to move slowly south in the direction of Galveston Bay. Even though flames were no longer coming out of the object, it still glowed brightly like a piece of a red-hot iron on a blacksmith's anvil.

As the object rose and began to move away from them, Betty blurted out: "Vickie, there's helicopters chasing that thing!" All three had heard the distinctive chop-chop sound of the helicopter rotors a minute or two earlier, but they had thought the sound was part of the noise coming from the object.

Although there were a few helicopters with only a single rotor, most of them had two large rotors, one in front and one in back. The helicopters were much bigger than the car, but they appeared to be small alongside the object. Betty and Vickie had never seen helicopters as large as these or with two rotors before.

The sudden upward movement left the helicopters behind momentarily, but they caught up with the object in a second or two, staying with it as it began to move slowly away.

"I'm going while the going's good," Betty said as soon as she realized the object no longer stood in their way.

But she realized for the first time that the engine on the Cutlass had died. The glare from the object was still bright enough that she had to feel for the keys in the ignition. She had to try twice before the engine started. As soon as it started, she shifted gears and jammed her foot down on the accelerator. As the car leaped forward, she turned on the air conditioner.

The object and the helicopters were off to their right but still not very far away. Some of the helicopters seemed to be swarming around the object while others were still flying toward it. "It looks like to me they're trying to help it or hem it in," said Betty. "See how they are stacked up around it? It looks like some of them are trying to get above it." Betty was driving south, and the helicopters were moving at an angle to the southwest. She had thought for a second about turning around and going the opposite direction, rejecting that idea in favor of getting home quickly.

Original sketch of incident made by Kathy Schuessler from the witness interviews.

CHAPTER 3
The Trip Home

All three were still badly frightened, but the terror that had gripped them was now easing. Both Betty and Vickie were still crying a little, even though they were relieved that they had survived their ordeal. Colby was also crying at times, but not as much as before.

About two miles down the road, Betty stopped near a small bridge and said, "Look, there's more helicopters coming over there."

She pointed to the left, and low in the sky; they could see a steady stream of choppers flying toward the object from the east. By now, the object was perhaps a quarter of a mile or more away, and it looked like an oblong ball, glowing red. They never saw any more flames coming out of it after it lifted up above the trees.

Betty drove on, watching the object and the helicopters almost as much as she watched the road.

"You better watch where you're going, or you're going to run us into the ditch!" Vickie warned Betty.

All three were hot and miserable.

"I'm burning up from the inside out," Betty complained. "I feel like I'm burned to a crisp."

"I wouldn't wonder," Vickie replied, thinking of how Betty had stayed outside the car so long. Then she added: "Now I'm beginning to get a headache."

Colby complained of a headache, too.

"It is probably because of that glare," Vickie said. "My eyes are burning too.

Betty's eyes felt as if she had stared at the sun too long. All three could see spots before their eyes.

As Betty drove, Vickie watched the helicopters for a moment and then said, "One thing is for sure, either they know what that thing is or they're trying to find out."

"They must know what it is, or they wouldn't be there," Betty replied.

Colby was silent, keeping a fearful watch on the glowing object. "Those people in the helicopters better watch themselves," Betty said. "As hot as we were back there, they've got to feel it too. Some of them are even closer to it than we were, and if they don't watch out, they're going to get burned up."

"Maybe they don't know how hot it is around it," Vickie said. "I hope they don't get hurt."

Original sketch of the encounter made by Vickie and Colby Landrum drawn in March of 1981

Two miles further down the road, they turned west on FM2100 and a short distance later, Betty stopped near the entrance to a rural cemetery. From there, they could clearly see the object and the helicopters across an open field and had no intention of getting close enough to be burned again.

While they were stopped, they began counting the helicopters. Betty thought there were 26. Vickie and Colby were counting together, and Vickie said there were twenty-one. "No, there's two more over there, Grandma," said Colby.

"Maybe I counted some twice," Betty said. They finally agreed there were at least 23 helicopters, some flying around the object and others still flying toward it from the east.

As they watched the activity in front of them, one of the large helicopters with two rotors flew fairly low over their car. The noise and vibration were horrendous, instilling a bone-chilling fear into each of them. There was no indication that the crew members had seen them or the car.

The helicopter was so close to their car that they could see some kind of round insignia on the side that indicated to them it was a U.S. military helicopter, but they couldn't tell whether it belonged to the Army, the Air Force, or what.

"As soon as I get home, I'm going to call Ellington and ask them what is going on," Vickie said. At that time, Ellington was an Air Force Base located just south of Houston, about forty-five miles from where they were at the moment. Today, Ellington is a public airport owned by the City of Houston. A number of aircraft are assigned to the base, some of which are used by astronauts and other National Aeronautics and Space Administration personnel who work at the Lyndon B. Johnson Space Center a few miles from Ellington. Eight CH-47 Chinook helicopters, which are large and have two rotors, were also based there. The helicopters belonged to the Texas Air National Guard. The women did not know this at the time.

Betty turned east on the Huffman-Eastgate Road about a mile beyond the cemetery, which led them to FM1960, about two and one-half miles beyond. The object and the helicopters were now much farther off to their right but clearly visible.

At the intersection of the Huffman-Eastgate Road and FM1960, they stopped to re-count the helicopters. Again, they counted 23, but they could see still more coming from the east.

"Listen," Vickie said, "Let's don't stop anymore and take time counting the helicopters. Let's get out of here."

Betty agreed and turned the car onto FM1960, heading for Dayton and home.

About a minute later, the radio began playing. No one had touched it. Betty had turned it on when they left New Caney, but it had been silent since the first moments of the encounter. From that moment until now, they realized they had heard nothing from it, not even static.

Betty dialed one station after another, hoping to hear news about the object. But none of the stations said anything about it. They were now moving directly away from the object, and Betty could still see it sometimes in her rearview mirror. Vickie kept looking back from time to time all the way to Dayton while Colby, quiet now, was facing backward, watching out the rear window, still afraid the object was going to come after them.

"We never did see Jesus, did we, Grandma?" he said.

"No, baby, we didn't," she replied.

The closer they got to Dayton, the worse they felt. Vickie said she was feeling a pain between her eyes, and Betty complained of a severe headache. "It feels like the top of my head is coming off," Betty exclaimed.

Their eyes began to feel puffy and irritated. They were quite thirsty. They were beginning to feel nauseated, with abdominal cramps.

By the time they entered Dayton, Betty was so sick that she was barely conscious of letting Vickie and Colby out at their house. Any other time she would have stopped in and talked for a while. "Vickie, I'm just going to go on home," she said. "I'm so upset."

She sped off as they were still walking toward their house. As she drove the seven miles to her home north of Dayton, she could feel some lumps forming on her neck and head. However, the pain in her head was so great that she was only mildly curious as to what the lumps could be. She wanted only to get home as soon as possible.

"I'm going to get inside and lock my door," she told herself. She hoped that as soon as she had a chance to get a drink of water and lie down, she would feel all right.

"I feel like I could die," she said as she walked into her house. It was nearly her last lucid moment for about a week, and in the months to come, there were times she wished she would die.

CHAPTER 4
The Sickness Begins

● ●

Betty, a trim, attractive woman who was meticulous in her appearance, found the ravaging effects of the encounter on the highway were already beginning to show even before she reached home.

Betty Cash, 1979, as she appeared before the incident.

Less than ten minutes after letting Vickie and Colby out, she walked into the house feeling nauseated. Her face was red and swelling, her head was pounding from a headache, and her eyes were burning. She was unbelievably thirsty.

She was also afraid. Traces of the terror she had felt less than an hour earlier remained, and she wanted nothing more than to get inside, lock the doors and go to bed.

But she found she had company. Wilma Emert, 42, a friend who worked for her as a waitress in the small cafe that Betty operated, was there with her son, Darrin, 15, and a niece, Leslie, 10. Wilma's husband was working the night shift, and she sometimes stayed at Betty's overnight rather than going to her home far out in the country. Betty had forgotten that Wilma was coming over.

Betty's son, Toby Howard, 32, was also there, but he didn't stay long, and

Betty has no recollection of seeing him.

"Wilma," Betty said, trying to focus her blurry eyes, "I'm so sick, and I've never had a headache as I've got now. It's really terrible. And look."

Betty passed her fingers lightly over some of the lumps that had formed and said: "Look, they're all over my head, my neck, my arms."

"What's happened to you?" Wilma asked. Betty told them about the incident on the highway, but Wilma didn't think she was being serious and just laughed.

The nightly 10 o'clock television news was coming on and Betty tried to watch to see if there was anything about the errant aircraft, but her eyes hurt too much, and she felt as if a veil had been drawn across them.

Shortly afterward, Betty said good night and went to bed. By then, she had drunk one glass of water after another, yet she still felt thirsty. She was also beginning to vomit and suffer from diarrhea.

A somewhat similar scene was being played out in Vickie's home in Dayton. For Vickie, a usually cheerful and alert woman with pale blue eyes and gray hair, it was "the longest night of my life, a nightmare."

She and Colby both felt as if they had been out in the sun at the beach all day, but because the weather had been cool and windy, she thought they had simply been chapped.

Both drank several glasses of water apiece, trying to quench their thirst. Vickie then took a bath in an attempt to relieve the burning feeling, but that only made her feel worse. The water, although tepid, felt scalding hot.

Because of this, she decided to just sponge Colby off with a damp washcloth. Even that was too much, and he complained that it was burning his skin. When she finished, she rubbed baby oil all over and put him to bed, and then rubbed baby oil on herself. Vickie had trouble going to sleep. She was feeling miserable, and she kept thinking about what had happened to them. She finally fell asleep, only to awaken about one a.m. to hear Colby crying. She went into his bedroom and found he had vomited all over the bed. He was also having diarrhea, and as Vickie began to clean up, she got sick herself.

She couldn't understand why Colby was so sick. She thought perhaps it was something he had eaten at the truck stop in New Caney, but all he had was some pancakes and milk. It never occurred to her that his condition, and hers, could have been connected in any way with their encounter on the highway.

Both Vickie and Colby experienced skin irritation following the event. Avon donated a box of skin care products that they found effective, delivered by John Schuessler.

She gave him some Kaopectate and managed to get him back to sleep, but neither of them got much rest. Throughout the night, she and Colby both continued to vomit from time to time and suffer from diarrhea.

Betty was having even worse problems. She was desperate for water but was too weak to get out of bed. She felt like she was on fire inside. Several times during the night, she called out to Wilma for help, but she couldn't arouse anyone.

Betty's home in Dayton, Texas

The next morning, not aware that Betty was sick, Wilma and the children left early. Betty, dozing fitfully at times, didn't hear them leave.

Back in the Landrum home, Vickie had gotten up at 5:30 a.m., still vomiting and with an aching head, to fix breakfast for her husband, Earnest. He had been asleep when she and Colby got home the night before and had slept through all the commotion. Vickie's skin was redder than usual, but he didn't seem to notice. Vickie didn't call his attention to it, nor did she say anything to him about what had happened.

Sometime after he left for work, she awakened Colby, who was still sick and had him get dressed. By now, both of them looked quite sunburned, even on their arms although they had been wearing sleeves the night before.

"Colby," she said to him, "don't tell anybody about what happened last night because they won't believe you." Neither felt much like eating breakfast and, around 9:00 a.m., she and Colby drove to Betty's restaurant and grocery, located seven miles north of Dayton. Vickie worked there as a waitress, but Betty had closed the business down a few days earlier and was preparing to move to a larger location in Dayton. Vickie had promised to help pack, but Betty hadn't arrived yet, so Vickie used her own key to unlock the place. She then sent Colby over to Betty's house about a third of a mile away to see why she was missing.

Colby found the doors locked, so he pulled himself up to the. windowsill to peer inside. Colby came back and reported there was no activity in Betty's house.

Vickie worked all her life, even while raising her five children. A confident, self-reliant woman, she grew up working in the cotton

: *Colby demonstrates how he looked for Betty in her house when there was no response.*

fields near Laurel, Mississippi, and later had worked in Dayton for many years as a waitress. Without waiting for Betty, Vickie started right to work. Loyce Edwards, the cook, arrived soon after and pitched in.

An hour had passed without any word from Betty, so Vickie asked

Edwards if he would go over to Betty's home and check on her. Edwards went there and knocked on the front door but got no response. He then went around to the back of the house, knocked on Betty's bedroom window, and asked if she was all right. Betty told him to tell Vickie she was sick and to come to help her.

Vickie found Betty unable to do anything for herself. Her face and neck were swollen, and she appeared to be covered with blisters.

"Vickie, I've got to have some water," Betty moaned.

"Betty, what's wrong with you?"

"I'm dying. I've got to have a drink of water."

Vickie got her water and helped her into the bathroom. She tried to get Betty to go to a doctor or even go to a hospital, but Betty refused. She thought people would laugh at her if she told them what had happened to them the night before.

Vickie tried to get Betty to eat something, but Betty had no appetite. Vickie stayed with her for several hours, bringing her water and helping to the bathroom from time to time, but by early afternoon she realized Betty wasn't going to get well enough to be left alone. She bundled her up in a blanket, helped her into her car, and took her back to her own home.

Vickie fixed up a daybed in her living room for Betty, who continued to suffer from vomiting and diarrhea, as did Vickie and Colby.

Late in the afternoon, Ernest Landrum, a big, husky fellow of 58, came home from his job at a petroleum tank farm and asked why she and Colby were so red. Vickie said she guessed that they had gotten chapped by the wind out at the cafe that morning. He then asked what was wrong with Betty, and Vickie, because of the blisters, said she thought Betty had blood clots.

Vickie nursed Betty for the next three days. Betty was delirious much of the time, and she ate very little. Vickie would try to feed her a spoonful of soup, but Betty would throw it up right away. Her skin was very red, and she was swelling. She was still suffering from a severe headache as well as nausea, diarrhea, and blisters on her neck, head, and back. Her eyes had swollen shut.

Vickie became more and more worried about her; afraid she was going to die. She got so scared she asked Loyce Edwards and Wilma Emert to come and look at Betty. They thought she needed to be in a hospital.

Neither Vickie nor Betty had told anyone other than Wilma about the encounter and Vickie had told Colby not to say anything about it to anyone.

To Vickie, it was "one big nightmare that wouldn't end." Betty continued to get worse, and Vickie didn't know what to do. She couldn't get a doctor to come to the house and look at Betty. For a long time, Betty wouldn't or couldn't tell her who her doctor was. When she did say her doctor's name was Dr. Shenoy, Vickie wasn't sure it was right. Vickie then realized that the local pharmacist might have the name of Betty's doctor and he did. It was Dr. Shenoy. She phoned the doctor, described Betty's condition, and he told her to get to Parkway General Hospital in Houston immediately.

CHAPTER 5
Betty Is Hospitalized

Vickie didn't have the money to call an ambulance, so she had her daughter-in-law help her drive Betty to the hospital. Betty was in so much pain she doesn't remember the trip. Betty was admitted to Parkway General Hospital in Houston on January 2, 1981, at two o'clock in the afternoon by Dr. V. B. Shenoy.

Vickie told the doctors she thought Betty was suffering from blood clots but said nothing of the encounter. The doctors quickly determined Betty didn't have blood clots and tried to determine if Betty had been burned. Thinking they were talking about being burned by fire, Betty told them no. She also said nothing about the encounter.

Despite her denial, they treated Betty like a burn patient, giving her antibiotics and steroids and applying salve and creams to the blisters. A dermatologist examined her and suggested she might have an allergic reaction to something, possibly shampoo. A neurologist examined her and said her headache might have been caused by severe tension. X-rays, and EEC, and a CAT Scan showed no abnormalities.

On the second day in the hospital, Betty began losing hair, and over the next couple of weeks, she developed bald spots all over her head. Her eyes swelled up, and she could barely see. She kept breaking out in blisters. The headache continued, as did the bouts of nausea and diarrhea. Whenever she closed her eyes, she could still see the glare from the encounter with the object.

Two days after Betty was admitted to the hospital, she was visited by her daughter, Mickey Foster, 31, from Dallas. "I walked through the door and turned right around and walked out because I thought I was in the wrong room," Mickey said later. "I didn't even know it was my mother. I had to walk out and look at the door number again. It upset me so badly I broke down crying. Her face was the size of two man's faces put together. Her eyes were swollen shut. It looked like she'd been under a sun lamp. When I rubbed her forehead, water just squirted out. It looked like a blister that had popped, but it was all over her face, and it was all in her hair. I went to rub her hair, and it was just falling out. She didn't know I was there until I spoke, and I'd been sitting there for ten minutes trying to get my head together."

On January 19, Betty was allowed to go home, but the doctor warned Vickie, "Don't let this lady lay home and be in the shape she was in before you bring her back."

Betty didn't fare well outside the hospital. The horrible headaches continued. She began to swell again and broke out in more blisters. She had no appetite, was sick to her stomach, and would throw up even if she took just a drink of water. So, on January 25, she re-entered Parkway General Hospital for more treatment and more tests.

It was when Betty was being admitted the second time that the story finally came out. One of the doctors asked what had caused Betty's condition, and when Vickie and Betty said they didn't know, Colby spoke up and said: "I know what it was— it was that thing we saw." When the doctor asked for an explanation, Betty decided to go ahead and tell the rest of the story.

Dr. Shenoy, Betty's regular doctor, was upset when he learned of the encounter. He chastised her for not telling him the first time she was admitted, saying: "There's a chance I could have helped you rest a lot better if I had known." He then started her on antibiotics to help fight the infections in the blisters. Her eyes were still swollen, and her ears were infected.

But the real story was not of much use in determining how to treat her because the doctors were really at a loss as to what to do. "I have never had any experience in dealing with a patient who has been exposed to a UFO," Dr. Shenoy explained later. Although some doctors sniggered at her story, they were fortunate he took it seriously. After Betty was admitted the second time, the loss of hair became much worse. It fell out in clumps. In a few days, she had several bald patches on her skull three to four inches in diameter. Despite the story about the UFO encounter, a dermatologist said the problem was alopecia areata, a baldness disorder that may or may not be permanent. Later investigations revealed this diagnosis was suspect.

An opthalmologist examined her eyes to see what effect exposure to the bright light might have had but found no severe damage. Results of blood tests were never released. The EEC and CAT Scans revealed nothing of interest.

She was discharged for the second time on February 9 and stayed with a brother in Houston. Two weeks later, her mother, Pauline Collins, took her back to her home in Birmingham, Alabama, so that she could look after her.

Betty never regained the health she had before the encounter. She had lost 16 pounds, dropping to 94 pounds, and it was months before she was back up to her usual 110 pounds again. It was four months before all the vomiting, diarrhea, and nausea stopped, but the bad headaches continued for six months. It was May before her hair began to grow back, and then it took several more months before she finally gave up wearing wigs when leaving the house.

For an exceedingly long period of time, she would break out in blisters when she was near any heat higher than normal. That meant she was no longer able to take warm baths and had to bathe in water that was nearly cold. Otherwise, the water felt like it was burning her, and blisters would start forming in about 15 minutes.

Betty entered a hospital in Birmingham in late May because of skin problems and was discharged four days later. She entered the hospital again on September 29, complaining of chest pains. The diagnosis was bronchitis, and before she was well again, she asked to be discharged on October 4 because an aunt was dying. She was re-admitted on October 17 and stayed another ten days. One month later, on Thanksgiving Day, she was again admitted because the chest pains were back.

During this visit, a dermatologist examined her, and he said she was suffering from "radiation exposure."

On February 13, 1982, she received emergency treatment after falling in the bathtub and injuring her elbow.

Chest X-rays taken throughout 1981 and 1982 showed no abnormalities, but early in 1983, lumps were found on her right breast. The breast was removed on March 29, and Betty began chemotherapy treatment. In May 1983, she suffered a heart attack and on June 23, her other breast was removed.

The pattern of hospital visits has continued every year since the encounter and are still continuing. While doctors cannot say that the encounter on

December 29, 1980, caused all of the events, they do admit that it could have been the "trigger" for the events that followed. There is no doubt that she was exposed to radiation during the encounter and that the encounter left her in a weakened condition that lowered her resistance to illness and disease.

[Editor's Note: Betty Joyce Collins Cash passed away on December 29th, 1998 at the age of 69. This date marked the 18th anniversary of the event.

Betty Cash's written description of the UFO event for the Parkway Hospital Records.

> **PARKWAY HOSPITAL**
> 233 W. PARKER ROAD – 697-2831 – HOUSTON, TEXAS 77022
>
> was so hot we had to turn on the Air Conditioning. We got to the stop Sign on 1960 + I Counted 23 Helicopters. As I sat there looking at the object, I carried Vickie + Colby Ronne let them out + it took me about 20 minutes to get home. By this time I had a Terrible Headache. Lumps were all over my neck + head, my face was Red like it had been burned + my Eyes were swollen Shut + later my Eye Lids Bursted after I was in the Hospital. My face Pilled off + most of my Hair is out + it is still coming out. I was Desperatly Sick within 30 minutes after looking at the object. I stayed home trying to get better until the 2nd Day of Jan. 1981 And I'm still here on Feb 6th 81. I went home for 9 days But I was so sick again with the Headaches + Lumps on my head I had to return to the Hospital and I'm still here. I know I was not sick until this happened + I know that is what is wrong with me.

Second page of Betty's report.

CHAPTER 6
Vickie Toughs It Out

Vickie Landrum is a tough-minded, self-reliant woman who would have survived in the early pioneering days of the old west, but nothing had prepared her for the ordeal she went through following the encounter. She grew up working in the cotton fields near Laurel, Mississippi and married her childhood sweetheart, Ernest W. Landrum, when she was 16. She quit school after finishing the eighth grade to start raising a family. She and her husband have five children, all grown now with kids of their own, and they have lived in or near Dayton since 1948. Vickie worked for many years as a waitress in Dayton, while her husband worked at petroleum tank farms in the Houston area.

"It was like one big nightmare that wouldn't end," she says, looking back at the first few days after the encounter. I sat up with Betty and Colby for three nights, and I would have to stay up in the daytime to take care of them. I thought she was dying. I thought I was going crazy. You couldn't tell nobody because it was too unbelievable. She was sick herself with headaches, nausea, vomiting, and all the rest.

"For about three months, I'd wake up at night so sick to my stomach it wasn't funny. One night I went into the bathroom, and then I had to lay down on the floor to get better so I could go back to bed. In the daytime, it would ease off, and then I'd be sick again at night."

The headaches, vomiting, nausea, and diarrhea, along with the thirst and lack of appetite, continued off and on for about two months, but by then a new problem had developed. She lost most of her hair, just as Betty had. It was not until late in 1981 before most of it had grown back again. Vickie was pleased with her new hair because it had a different, coarser texture and was easier to manage.

She also broke out in blisters on her arms and hands shortly after the incident, and there was never a time during the following year that she didn't have new blisters forming or old ones healing. At the same time, she found that she had become highly sensitive to the sun and heat, such as from a kitchen stove or a heater. This problem continues to plague her, and even brief exposure can cause her to break out in blisters.

With the hair and skin problems, there were many times when she hated to go out in public.

"I had all those blisters and sores; my eyes were running, hair falling out—it was so embarrassing. I had to go to the grocery store once, and this old lady, a woman I've known ever since I moved to Dayton, she says, "Honey, aren't you ashamed to be in a grocery store in that shape?" And I said, "What do you want me to do, starve?" Then, Vickie went outside and cried.

"Another time, I went to church, and one of the old biddies said, 'Aren't you ashamed to be out in public like that?'" Vickie didn't reply. Instead, she got up and left. She said, "It wasn't very funny walking around with a bald head."

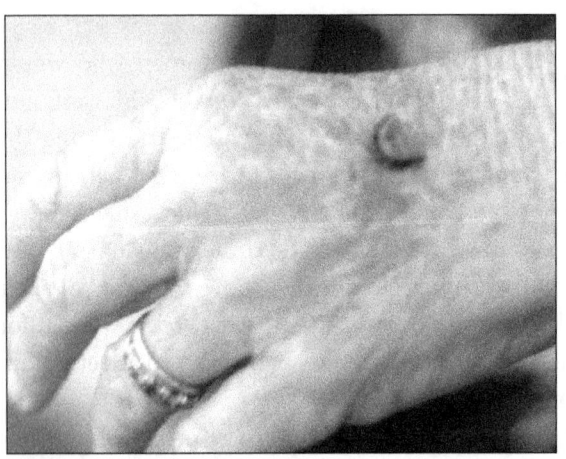

Vickie's hand, showing the type of sores present following the incident.

She tried wearing a wig, but this led to a peculiar reaction. Martha Thompson, a long-time friend, tells this story: "She has the wig on one night at my house, and she looked so nice in it, and the next time I saw her she had the wig off, and three or four more times after that I saw her with no wig on, and I said, "Why don't you wear the wig anymore?"

And she said, "Oh, that dammed thing. Every time I wear it, I can taste it. I can TASTE it! I can't explain it, but when I put that wig on, I can taste it in my mouth."

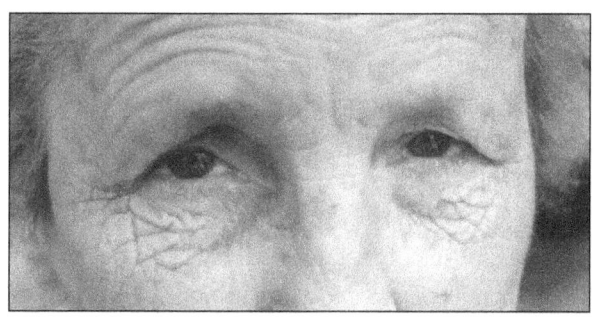
Eye inflammation after the incident.

Vickie's eyesight began to deteriorate. "My eyes just got worse and worse, and they were tearing, so I couldn't see, even at night. My pillow would be wet where they would run. In the mornings when I got up, it felt like my eyelids are stuck on something."

An optometrist in the nearby town of Liberty examined her eyes and asked her if she used a microwave oven. She said no. He didn't explain why he asked, but he did give her some medication which helped some. Another examination in April 1981 showed she was developing a cataract.

Except for the eye doctor, she didn't see any physicians. She and her husband had no medical insurance and, with her no longer working, doctors were a luxury they couldn't afford. "I just doctored myself," she said.

For the most part, she also doctored Colby through most of his problems. Although he is a grandson, he has been more like a son most of his life. When Colby was just a toddler, Vickie stopped working to raise him. Colby's parents, Paul and Peggy Sue Landrum were divorced when he was a baby, and later, when he was four, Vickie went to court and gained legal custody of him. Colby has lived with her ever since, but he frequently visits his mother and father.

Colby was in the second grade when the encounter on FM1485 took place. He hadn't wet the bed since he was a year old, but for about two months after the encounter, Vickie had to get up two and three times a night to change his bed.

Worse, his lack of control over his bowels forced her to put diapers on him for almost two weeks. Even though he was seven years old, he didn't complain—but he would have been mortified if his schoolmates had found out.

The diarrhea problem recurred from time to time for nearly a year. It would often cause him to have an upset stomach, and this would force him to miss school, once for an entire week and at another time for a couple of days.

"I'd give him Kaopectate, I'd give him Pepto-Bismol, and I'd give him some medicine the doctors gave me that smelled like it had paregoric in it," Vickie said. "I was going to carry him to the doctor, but when it got to where the doctors weren't doing anything for Betty, I said why to put Colby and me through it. If they couldn't find out what was wrong with her, they sure couldn't find out what was wrong with us, and we got along just as well as Betty did."

For more than a month after the encounter, Colby had very little appetite, and his thirst seemed unquenchable for almost five months. Colby didn't break out in blisters the way Betty and Vickie did immediately after the encounter, but he did become sensitive to the sun. In March 1981, he and his mother Peggy Sue went fishing on Lake Livingston, and although the sun didn't seem very hot, "when he came home his left cheek was nothing but a big hanging blister," Vickie says. After that, she took care to see that he didn't stay in the sun too long. His eyes were irritated for many weeks after the encounter, and he complained at times, saying, "It's like I'm looking through plastic." Just before school started in the fall of 1981, he had to be fitted for his first pair of glasses.

Colby Landrum at the time of the incident.

The psychological effect on him was perhaps worse than the physical. There were times when something as simple as a helicopter flying overhead could send him into hysterics. It was nearly six months before he would go outside and play after dark, even though the family lived on a well-lighted street.

On the night of March 7, 1981, he was in a car with Vickie, Betty, and another woman when he spotted what turned out to be an airplane with a bright light some distance behind the car. "He liked to have had a fit," Vickie said. "We had to explain to him that it was just an airplane."

"On another night, he and some other kids were playing outside, and they saw this star fall. He came screaming into the house, and I had to shake him to his senses."

Colby was also sometimes afraid to play outside in the daytime, usually after he had seen a helicopter or something similar that would re-awaken his fears. For example, Vickie says, in January 1982, "he saw two big helicopters like we saw, and for about two weeks he wouldn't do his schoolwork just so the teacher would make him stay in at recess. Every time he sees something, it puts him back into this slump."

[Editor's note: Vickie Marzelia Holifield Landrum passed away on September 12, 2007, at the age of 83]

CHAPTER 7
The Investigation Begins

It isn't every day that someone is confronted by a diamond-shaped object as large as a water tower that comes down out of the sky, blocks the highway for 15 minutes, and then lifts and flies slowly out of sight, trailed by a large number of helicopters. In fact, as far as can be ascertained, it has never happened before, not in the United States or anywhere else in the world.

It was such an extraordinary event that it should have been page one news the next day. But it wasn't - for two reasons. The first is that the news media never heard about it until nearly two months later. Neither Betty nor Vickie told anyone about the incident other than a few close friends and family members, and Colby had heeded his grandmother's warning not to talk about it.

The second reason is that most news people are too skeptical to believe such a thing could ever happen, and they would have rejected such a story out of hand without ever trying to find out whether there was any truth to it.

Betty and Vickie didn't know what the object was—an experimental aircraft, a "flying saucer" from who knows where, or some other device. So, it was called an unidentified flying object, better known as a UFO. This simply means that after examining and eliminating all possibilities, it remains unidentified.

But the term "UFO" tends to be repulsive to otherwise fair and objective news people. They lump it in with ghosts, bigfoot, the Loch Ness monster, tarot card reading, astrology, and other things they find disturbing or distasteful. They associate UFOs with "little green men from outer space," a term that is tossed off with sarcasm and ridicule.

Too few editors and reporters ever made any effort to take a clear, unbiased look at this unusual phenomenon. It is as if they had made up their minds that such things could not be, and therefore they aren't—despite the fact that over the past fifty years, hundreds of thousands of people around the world have reported seeing unidentified flying objects. And people still do. Virtually every day, somewhere in the world, including the United States, people see flying objects that cannot be explained.

Fortunately, this closed-minded attitude toward the phenomenon is not prevalent in many smaller cities, where reporters and editors tend to be younger and more willing to accept the possibility that such things might exist.

It was not until late February 1981— seven weeks after the incident— that a newspaper reporter in Conroe, a small city just north of Houston, heard about the incident and began checking into it. Although the women had told only a few people about the encounter, they did try to find out what the object was and why it had caused them to be sick.

While Betty was in the hospital, Vickie had phoned several police agencies, several radio and TV stations, and a newspaper, seeking information about the object they had encountered. None were able to help. Eventually, she went to see Tommy Waring, a neighbor, and the police chief in Dayton.

"We run into this thing that almost came down, and it burned us bad," she told Waring, showing him her reddened skin and telling him about Betty and Colby. "I think Betty is going to die." Chief Waring wanted to know what kind of "thing" had come down out of the sky.

"It was some kind of object," she replied.

Waring didn't know what to think about her story, but he had known her for many years, and he had never known her to lie to him.

"Well," he finally said, "I've got a UFO number somewhere in the office. I'll hunt it down and give it to you."

Vickie didn't think it was a UFO and said, "Mr. Waring, I think it was something the government had up there, and it liked to have fallen on us."

A short while later, Waring gave her the phone number of the UFO Reporting Center, a privately funded operation in Seattle, Washington, run by a retired fireman named Robert Gribble.

[Editor's note: NUFORC was founded in 1974 by Robert J. Gribble and is currently run by director Peter Davenport. The National UFO Reporting

Center has cataloged almost 90,000 reported UFO sightings throughout its history, mostly in the US. Nuforc.org] The Center's number has been distributed to police stations, sheriff's offices, and other places throughout the country.

Gribble and others take incoming calls and, if they sound legitimate, question the callers to obtain details of the sightings. They then pass this information on to one or another of the several national UFO organizations in the country.

These organizations have field investigators, members who work voluntarily in their free time and usually at their own expense, looking into reports of sightings.

At the time of the event, there were three prominent national UFO organizations. The oldest of the three, the Aerial Phenomena Research Organization (APRO), based in Tucson, Arizona, is now defunct. The Mutual UFO Network (MUFON), now headquartered in Ohio, is the largest of the three, now boasting more than 5,000 members. The last, the J. Allen Hynek Center for UFO Studies (CUFOS), based in Evanston, Illinois, is basically a research organization.

In addition to these three major UFO organizations, there are a number of small local or regional groups, such as the Vehicle Internal Systems Investigative Team (VISIT), located in the Clear Lake area of Houston.

I first learned of the case when I received a telephone call on January 26 from Howard Sussman, a Houston physician and personal friend. Howard explained that he had spoken with three of Betty's doctors and, without disclosing Betty's name because of medical ethics, provided me with an overview of the case. He provided an update on Betty's condition on January 30, again on February 2, and every day or so after that until Betty left the hospital. The official notification came after Betty was discharged from Parkway General on February 9. On the 16th, she telephoned the NASA Public Affairs Office at the Johnson Space Center near Houston. She said she wanted to find out who owned the helicopters and whether or not NASA knew what the object was.

She was assured that NASA owned no helicopters and did not know what the object was. However, since it sounded like she was talking about a UFO sighting, she was referred to me, as I was known in the area to be interested in UFOs.

She phoned my office the same day. I was not in, and Betty left a message with my secretary. Over a period of several weeks prior to that, I had received a number of calls about UFO sightings from people who turned out to be not too reliable and, thinking Betty's report was another false alarm, I didn't place a high priority on returning her call.

It was on February 17 that Vickie first phoned Cribble at the UFO Reporting Center. He mailed the incident reports to APRO and MUFON, and both, in turn, relayed the information to VISIT, the principal UFO group in the Houston area at the time. APRO's report was picked up from VISIT'S post office box on February 25, and MUFON's on March 1.

On February 21, Cathy Gordon, a reporter for the *Conroe Daily Courier*, phoned me to discuss the case, having obtained my name and phone number from APRO's headquarters. She was concerned about the women's health, and she stressed the serious nature of the case. The next day, February 21, Betty called me at my home. Having been briefed on the incident by Cathy Gordon and concluding it was the same incident described earlier by Dr. Sussman, I made an appointment to interview Betty the next day at her brother's apartment in Houston. This is where she and her mother were staying since Betty left the hospital.

One look, and it was clear that Betty had gone through a very traumatic experience and was slowly recovering. Her appearance—her face was still swollen, she had blisters on her arms and face, and there were large bald patches on her head—shocked me and helped to convince me that she might be telling me the truth.

I promptly phoned Vickie Landrum and made an appointment to interview her, and several days later talked with Betty's physician at the hospital. That is how the investigation began.

CHAPTER 8
The First Documented Report

On February 1, seven days into Betty's second hospital stay, she and Vickie each made an audio tape recording describing the ordeal. It was at the urging of friends and hospital workers alike that they borrowed a recorder and made the first documented report of the incident.

Kathy Schuessler transcribed the exact text of that recording as follows:

Statement by Betty Cash:

"This is Betty Cash. On December the 29th, approximately about 9:30, 9 to 9:30, a friend of mine and her grandson had been to Cleveland to try to play bingo. On our return home, we faced an object.

We came through New Caney, came out, went to eat. I never felt better. I had no illnesses, none whatever. We took a short cut back through to Dayton. It was the Huffman and New Caney road. And at this time, we had spotted something in the air that was long, that looked like the sky had just split open. In fact, I thought the world was coming to an end. It was bright, the lights were bright. And there was a lot of heat coming from this object.

We stopped and tried to see what it looked like. We could not get up close enough to detect what the figure was. Or I could not, at least, the lights were too bright in my eyesight. So, I stood with my arms on the window- on the door- and on the top of the car trying to see. But at this time, the heat was too intense, so I got back into the car.

I started the car up, and at about the same time, this object went up into the sky, and it was red and lit up all over. It looked like just the whole sky was open and lit. But there was quite a few helicopters circling around. I don't know whether they were trying to get around it or up closer to it or what, to see maybe what it was.

I got back into the car, and the car was so hot it was unreal. It was just like you had run it into an oven. When we got to the corner of 1960 and this Huffman-Crosby road, we had a stop sign. We stopped and waited for traffic—which was none or any that we were paying that much attention to. We were too busy watching the object.

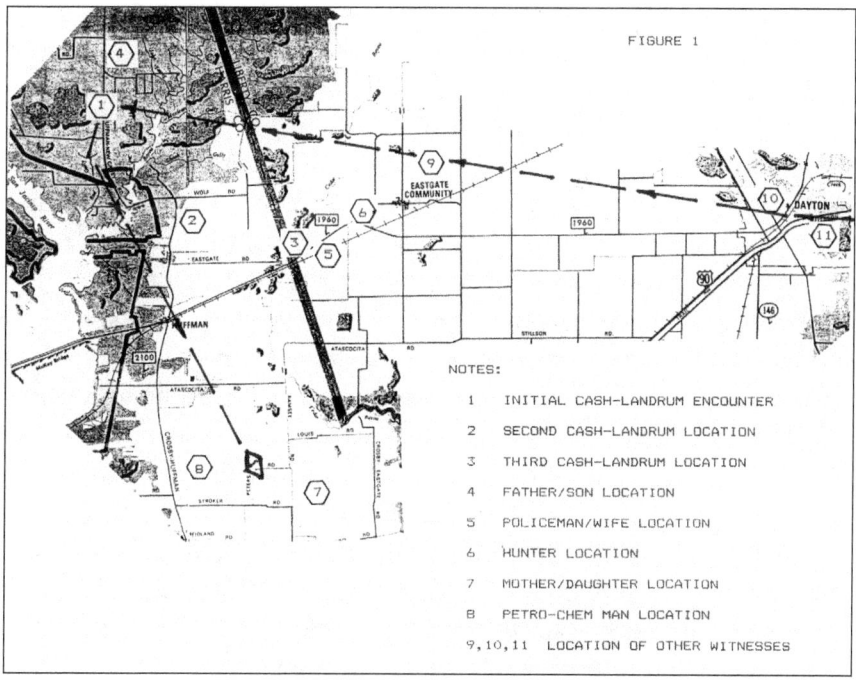

Map of the suggested flight path of the UFO.

But at this time, I counted 23 helicopters, around and about the object. They were far away, but yet they were low enough, and we set there and watched them 'til they got over the car because I wanted to make sure if it was airplanes or if it was helicopters, which it was helicopters. I counted 23 of them. I don't know what color they were, I can't say. But I do know that they had a double deal on the top, propeller-like thing. And I could hear 'em just as plain as if they were right ready to land, and at this time I turned left on 1960 and went into New Caney—I mean Dayton. I let Vickie and her grandson out and I went on home which took maybe 20 to 30 minutes. And I had big knots coming up on the back of my neck and on my head. I was just deathly sick. Well, I went to be thinking well, y'know, that I might be taking the flu. I had no idea what the cause of it was because I felt so good that day! But the next morning, I tried to get out of bed to get a drink of water, and I wasn't even able to get out of bed. Every time I touched my head, it was just like it was I was coming unglued.

Well, I stayed there 'til Vickie got off that afternoon. She came down and got me water and milk and things for the day. When she got home from work, she came down and picked me up and carried me to her house. And I stayed there four days before I went to the doctor. She finally made me say I would go to the doctor. And she called Dr. Shenoy and asked him if he would take me—if he would see me at the emergency room, that there was something terribly wrong with me. Well, he agreed.

She brought me to the emergency room. And at this time, I was swollen so bad, my ears had even looked like they were fixing to burst. Those knots on my head bursted, and they were just like blisters. Just like, uh, you had burned your head with something it would make a water blister.

Well, they checked me into the hospital. They said they didn't know what it was. And the little admission nurse asked me if I had, if I was a burn victim. And I told her no, that I just took sick, and I had no idea what it might be. So, they admitted me to the hospital, and I was there 12 days. At the end of the 12 days, the three doctors told me that they still had no earthly idea of what it might be. I went home.

I didn't feel good the entire time I was home. But I did stay one week, and I thought, well, maybe it would get better. At the end of that week, I was so deathly sick again that I had to come back into the hospital. In the meantime, from the time I entered the hospital until I re-entered the hospital, almost all of my hair had come out just in great big patches. It's just completely bald, practically. And I began, and the doctor asked me to use some Betadine for shampooing it, not to put any hair spray at all on it. Which I don't have enough to spray, and that is all that I have used since I was last dismissed from the hospital.

Well, today is my seventh day here, and they still can't tell me what is wrong. There, they've took X-rays. They've done everything I guess that they know to do for me. But yet, they can't come up with an answer of what's caused it. Well, I know what's caused it because I had never been in that predicament before. And I've never been that sick. So, l know exactly what caused.it. But of course, the doctors here can't seem to find it. They said they'd never seen anything like it.

My eyes are still burning. My vision is not clear. It, it feels as though I've got a skim over my eyes. I went to an eye doctor. He said that he could not tell if it was, uh, an overdose of radiation at this time or not. That it would be real, uh, a little early in order for him to tell, that it would take some time.

So, other than that, this is all I know, but I know for a fact that object, whatever it may be, is what has caused my illness. And if y'all have any information that could help me in any way, I would be most appreciative.

Original map sketch of the sighting sequence, drawn by John Schuessler during interview with witnesses.

Statement by Vickie Landrum:

"My name is Vickie Landrum. On December 29, 1980, Betty Cash, my seven-year-old son, Colby, and me was coming from New Caney by the Huffman and New Caney road. Betty was driving. When all at once, something came down. It looked like the whole sky split ahead of us. We stopped.

She got out of the car and stood for, I don't know how long, and my son and me go out on the other side and stayed for only a minute or two. For the little boy who has just turned seven was screaming and about to have a heart attack. So, I got back in the car and took him in my arms. I told him it might be Jesus coming after us. If he saw a man not to be afraid, He would be coming to carry us to a much better place. The whole road ahead and around it glowing as if by fire. I believe it was fire because it glowed down and let up a little. Or if it was, as if it was, it was coming from an old object up above.

Colby swore it looked like a big diamond. I couldn't tell for I was so scared about him. I didn't know how long it would be before it would all be over. It lifted, and I knew it was at least a half-mile or more across the main part of the light. It (the object) was bigger than a water tower. Inside, the car was so hot, I turned on the air conditioner when Betty got back in the car.

There was some helicopters up there then. But as it lifted where we could get by, we traveled on to Farm Road 2100. We stopped by the church and looked—we looked again to which way it was going. It veered ahead and to the right of us. There were helicopters up there.

We turned on 1960 and Huffman road. When we got to 1960, we stopped again and counted 20 or 25 helicopters up there. I could have, it could have been less or more. I was so upset it really didn't matter, for I didn't care. The helicopters had two deals on top in place of one.

We got home, my eyes and face was burning like it had a bad sunburn. I know it wasn't make-up for I don't wear any. Colby's eyes and face was red like he had been in the sun. I put baby oil on both of us, for I figured it was the wind outside that had burned us. I hope and pray that is all.

Betty is in terrible shape. And I, as has been since that night. I tried four days to get a doctor to see her. But being the holidays, no one would see her, for she is a heart patient, and no doctor here or in Liberty would touch her. The morning, the fourth morning, she looked so bad, and her face would swell, and her head was hurting so bad she was out of her mind. I talked to her, and I said Betty, are there some doctors you can tell me about? She told me about her doctor, which I called, and it was Dr. Shenoy. And I carried her to the emergency room, where she is now. I thought it was blood clots or something. I didn't believe she would live 'till I got her to the doctor.

But doctors can take it from there. But somewhere, somehow, someway, maybe somebody can do something for her. For she's only a friend, and I'll do the best I can by her. But if somebody knows how to treat her or how to help her in some way, please God in heaven have mercy upon her and let her live."

CHAPTER 9
Betty Cash Is Interviewed

Although Betty Cash had provided many details of the case by telephone, I arranged to meet with her on Saturday, February 22, 1981, at her brother's apartment. Present also was her mother, Pauline Collins. Mrs. Collins had come to Houston to care for Betty and eventually take her back home to Birmingham, Alabama.

Betty had been resting before I arrived at the apartment, but she got up, donned a wig to cover her baldness, and proceeded with the interview as agreed. It was evident that she was still quite ill. She was weak and unsteady, and sores were plainly visible. And she said she had the same splitting headache that had been present every day since the accident. Before the interview was over, she showed me the prescriptions she was currently using and the listing of medicines she had used since her entry into the hospital.

Mrs. Collins was extremely concerned. She said she didn't recognize Betty when she arrived at the hospital for her first visit after the incident. She said she asked the nurse to point Betty out to her. Betty's face and neck were so swollen that no one would have recognized her. Even Betty's ears were swollen— swollen so much that her earring screws would not reach through the earlobes.

Betty was quite concerned about her appearance, so she quickly told about the blisters on her face and head. She said: "They'd pop up just like water blisters that you'd get if you had gone to the beach and laid out in the sun too long. And they had to put sun-type salve, the same as they put on burn victims, on my face to keep from leaving my face scarred. And they just pulled sheets of tissue off my face every time they'd apply that stuff. Boy, was I burned!"

Results of the question-and-answer session after Betty's opening statements are given verbatim in the following paragraphs:

(Q) What must you have thought during all that time?

(A) It seemed like the end of the world. I mean, that's the first thing that entered my mind when we first saw it. It looked like it just split the sky wide open, and here we was. And the whole sky lit up. Vickie looked at me, and she said: 'Well, Betty, this may be the end.' She was hoping that God thought she had lived a good life.

(Q) She really thought it was the end that night?

(A) Yes, and I did too. My gosh, I mean, who goes driving along a country road and expects to have something fall out of the sky, and there you are.

(Q) Where did you first see it?

(A) I saw it before Vickie did because the sky was so red and weird-looking. And I said: Vickie what in the world is this? And she said: "Betty, I don't know." We went around a curve and down a long straight stretch, and it was right there—right over the road.

(Q) What happened then?

(A) It was at treetop level, and it had flames shooting from the bottom. I thought about trying to go under it. I was just trying to think about how I could get out of there, really. There was no way we could back up on this old country road. You know how wide they are. And there were no turns to where we could turn in a driveway or anything.

(Q) How would you describe the flames?

(A) Well, you could hear it. I mean, it was like it was going SHEW! SHEW! SHEW! It sounded like a flame thrower.

(Q) During our telephone conversation, you mentioned a beeping sound. When did you hear that?

(A) The entire time, once I got the car stopped. I had the radio going. We were riding along, listening to the radio and yakking. We couldn't figure out whether the beeping was coming through the radio or if it was just coming from this deal. I still hear it.

(Q) You still hear the beeping?

(A) I mean, I don't think I will ever forget the sound. I won't forget because I go to bed at night trying to figure out how did it get from where it was when we first spotted it to there before we did. Or by the time we did.

Betty went on to relate how she had stopped the car in the middle of the road by coming to a sudden halt. Then she described how they got out of the car and stared at the object in front of them. She made it clear that Colby was terrified and how he tugged at his grandmother until she re-entered the car and held him. Then the questioning turned to the subject of how the object left the area.

(Q) What happened when the object left?

(A) When it left, it went up in the air, and it went toward Houston. I had my eyes focused on it as much as I could see. The light from that thing was unreal. It was like somebody was holding a real bright light in your eyes. And you can't really see too much. In fact, I had to sit there for a while to even be able to see the road to drive home. We were all blinded that way.

(Q) How did you find out how to call me after the incident?

(A) Well, Kathy Gordon is the one who said, well Betty, why don't you call down at NASA and see if they can tell you anything. So, I did. I got your work number out there, and I called you.

(Q) The NASA operator gave you, my number?

(A) Uh huh. And I said, well, I'm calling long distance, and I left my number for you to call back. Then we turned around and called Ellington (Air Force Base near Houston), and we couldn't get any information out of them at all.

(Q) On the telephone, you mentioned that your medical bills were quite high. How expensive were they?

(A) My hospitalization alone has been way over ten thousand dollars, and that doesn't include any of the doctor bills. That's just the hospital alone. And here I am. I can't work. So, what am I going to do?

(Q) Your insurance won't fully cover the expenses?

(A) No way!

(Q) What was the doctor's diagnosis?

(A) Well, he was just baffled. They ran all kinds of blood tests. They even got my glucose tolerance test, everything. They couldn't figure out what was making me swell, why I was burned, why my hair was coming out.

(Q) That is a weird combination of symptoms.

(A) And I didn't tell them about this weird ordeal I had gone through. So, I imagine he really was baffled because he'd make two or three trips a day to see me. He'd say: Miss Cash, I can't figure it out. I just don't know. I just don't have an answer for it. So, when I had to go back to the hospital the second time, and Vickie's grandson told him what had happened, he was a little perturbed with me for not telling him the first time. No telling how many hours sleep he had lost wondering about it.

Betty then provided copies of her itemized hospital bills, listing all of the tests they had conducted and medicines she had taken. The listing was extensive. At that point, she verified the statements about the cost of her treatment by showing me the cost numbers on the bills.

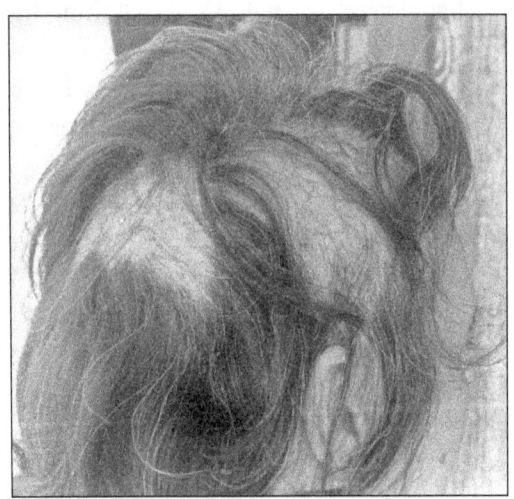

Betty's dramatic hair loss, in the weeks and months after the event.

She then removed the wig and allowed me to photograph her head from various angles. This was very embarrassing for her, so she kept talking about how nice her hair was before the incident.

She concluded by saying: "if and when my hair grows back, I'll show you a head of hair!"

The interview then moved outside the apartment to the apartment parking lot where Betty's car was parked. Betty walked through a role-play of the incident, starting by sitting in the car like she was stopping it in the road. Then she got out and stood by the door like she was watching the object. Then she walked to the front of the car, shielding her eyes with her arm, and gazed skyward. The role play concluded as she returned to the door, and finding it hot, used her coattail as a hot pad to grab the door handle and open the door. During the role-play, the whole sequence of events, starting with Betty stopping the car and getting out until she was safely back inside the car, took approximately eight minutes. She didn't know it at the time, but this long period of exposure was to cause her health problems for the rest of her life.

The car was a 1980 Oldsmobile Cutlass Supreme with Texas license number VAS 217. I examined it for obvious damage and found it to be clean and in good condition. The exterior paint and plastic parts were all found to be in good condition. The tires were like new. The only visible anomaly was some very clear hand-shaped imprints in the padded dashboard on

Betty's 1980 Oldsmobile Cutlass, the vehicle they were riding in during the incident.

the passenger (right) side. A Geiger counter was passed over every part of the vehicle, but no readings above background radiation level were found. Also, no unusual strong magnetic fields were found by using a hand-held compass as a detector. When it started up, the engine did run a little rough.

By this time, Betty had overextended herself and was about ready to collapse. I thanked her for being so candid with me, and we terminated the session.

CHAPTER 10
Vickie Tells Her Story

On Saturday, February 28, 1981, physicist Alan Holt and I drove to Dayton to meet Vickie and Colby and document their condition. At the time of the encounter, the Landrum's lived in a rented home at 506 West Clayton, just a few blocks from the center of Dayton.

When Vickie greeted us at her front door, the effects of her earlier injuries were still quite evident. Her face had a scarred appearance, her eyes were quite infected, some hair was still missing, especially along the right side of her head, and sores were visible on her arms and hands.

After providing our credentials and gaining permission to audiotape the discussions, we settled in Vickie's living room to hear her description of the incident.

When they first observed the object, she said it was high in the sky—only a "fire-like" light. However, it rapidly descended to treetop level above the road in front of them. Because it was shooting flames downward, they were afraid to drive under it and stopped the car on the highway. Soon, the car began to warm up, causing them to open the doors and stand outside the car, but within the open-door area

Although she smokes, Vickie is a non-drinker and quite religious. She thought this was the predicted end of the world and kept talking about it as she watched the spectacle going on in front of her. She said: "The Bible says the sky will split and in a rain of fire, Jesus will come." She expected that to happen as she watched. When it didn't happen, she felt she wasn't deserving and ready yet.

Vickie described the object as oblong in a vertical direction as it hovered over the road. She said it was very bright— it glowed with brightness. The whole area was lit up like daylight. While it was easy to see the pointed lower end of the object, the glow was so bright Vickie had trouble making out the shape of the top part. On the other hand, she was quite clear about fire coming out of the bottom of the object.

In response to Alan Holt's questioning, she talked about the flames coming out of the bottom. She said: "This fire was coming out of the bottom of it. And it wasn't just one little streak."

(Q) Several different streaks?

(A) It was, I mean, it was fire. It was bright. Now it (the fire) would let up, and then it would come back.

(Q) And when it would let out the fire, it would lower and raise. Is that what you are saying?

(A) Uh huh. It was like when the fire would let up, it (the object) would come back down, and then the fire would shoot closer to the ground, y'know.

(Q) How close did that fire come to you?

(A) Like I say, we could feel the heat from it, and it was just like you walked up close to a fire, and you know how bright that can be. Y'know, like you're burning brush that makes a real hot deal, well, that's how hot it felt. And it was really bright.

(Q) What colors were involved?

(A) The inside looked kinda dark and yellow. Around that it was pinkish, other than the fire itself. The fire was fire, like flames from a bonfire or something.

(Q) Would the fire have been directed down toward the road?

(A) Right. Sure, would have.

(Q) Did the object make any noise?

(A) As the object hung above the road, it roared like a hurricane. When the flames came down, they would make a loud "whooshing" sound.

(Q) How big did it look to you?

(A) Well, it was a lot bigger than a water tank. It was bigger. I'll say the inside of it looked from a distance like it was as big as a water tank or bigger. It would have to be a little bigger.

(Q) How did it leave the area?

(A) It started floating off to the right. It pitched up like that (making a motion with her hand). It was just like something reached down and picked it up. And it went up and off to the right.

(Q) Did the flames keep coming out?

(A) When it got up, it seemed like the flames stopped. Then it was a long white light. It was never a ground ball. It was always long.

At this point, Colby was invited to come into the room and describe what he saw. His description was similar to what had already been described, except for two important points. He said he could clearly see the pointed upper end of the object, making it diamond-shaped. In addition, he said he could see helicopters that appeared to be "trying to find out what the object was."

Vickie provided a copy of a drawing made by Colby that showed the object and the helicopters. Vickie said, "He swears there were helicopters up there then. He said he could see them when he was sitting in the front seat." She and Betty didn't notice the helicopters until the object rose and flew away, pursued by helicopters. Then they counted more than twenty in the air at one time.

Vickie told about how they immediately started the car and headed for Dayton as soon as the object cleared the road ahead of them. She claims they stopped twice more, once on FM2100 and once at the intersection of FM1960, where they watched the object and counted the helicopters.

Vickie went on to describe as least two types of helicopters flying around the object and trailing the object over an area of several miles. She responded to questions about the helicopters as follows:

(Q) Could you describe the helicopters?

(A) Yes, sir. Some of them had only one propeller, and some of them had two propellers.

(Q) So, there were two kinds of helicopters involved?

(A) Two kinds of helicopters. Sure were.

(Q) On those with two propellers, were both propellers flat the same way, or was one of the propellers turned on its side?

(A) Well, one of them was lower than the other one; but both were flat in the same way. They were both on top of the helicopter.

(Q) Did the helicopters have any kind of lights on them?

(A) Yes, sir. They had those blinking lights like helicopters do. You could see the lights. The helicopters were also lighted by the brightness of the object.

(Q) Could you see any markings?

(A) No. They were just dark-colored helicopters.

Vickie said she truly feels that this was not anything unnatural. She believes the United States government was transporting or escorting something dangerous through the area, so there were so many helicopters.

Vickie then talked about the injuries to herself and to Colby and allowed us to photograph the evidence. She described how she had laid her left hand on top of the ear as she stood in the open doorway and how that hand was severely burned. She believes that her right eye and the right side of her face were damaged more because of the way she was standing while keeping an eye on Colby.

She said: "Colby's face was like it was sunburned, but they hadn't been out in the sun." Because it had been a little windy that day, she thought maybe his face was chapped from the wind. However, when she gave him his bath, he was hot and red, like he had been on the beach all day. She then bathed herself and found she was in the same condition. She then covered him with baby oil and did the same for herself.

During the weeks that followed, the hair on the right side of her head came out. By the time of our visit, it was starting to grow back, but she said her scalp felt like "it was asleep."

The fingernails on Vickie's left hand clearly showed evidence of damage. Each nail had an indention, line-like, across the nail from side to side. The index finger had a hole through the nail at the line.

Vickie and Colby both had headaches, diarrhea, and nausea continuing up to the time of our visit. Their stomach aches were to persist for over a year. Problems with their eyes persisted much longer.

Colby had nightmares for several weeks. He would wet the bed because he was afraid to get up at night.

Vickie also gave a separate accounting of what had happened to Betty Cash during and after the incident.

Following the interviews, we requested Vickie and Colby to lead us to the location of the encounter on FM1485. They were afraid to return to the site, but they realized it was an important part of the investigation to go there to walk through the event as it had originally happened and consented to do it. While at the incident scene, we made a timeline of the events as Vickie tried to repeat everything she had done on December 29, 1980. Interestingly, although neither Vickie nor Betty had been back to the site since the incident, they both were able to take us to nearly the exact same location. These separate site visits verified the location of the incident for us.

The damage to the road caused by the heat from the object was obvious. The surface of the road appeared to have been melted and resolidified. The damaged area contrasted very well with the old, cracked road surface that runs for miles through the East Texas Piney Woods. The distance from the damaged area on the road to where they had stopped the car was approximately 135 feet. Photographs of the area, the road, and the vegetation were made during each visit.

CHAPTER 11
Colby Answers Questions

Seven-year-old Colby Landrum responded openly to questions, as follows:

(Q) What did you see?

(A) I don't know what 1 saw. (It was) a diamond shape. It come down to the top of the trees, and it was real bright, and the car got real hot. We had to turn on the air conditioner. And then Betty got out of the car and started walking towards it. I got out, and then I got right back in the car.

(Q) Why.

(A) Cause I don't want to get hurt as bad.

(Q) What did you think?

(A) I thought the world was coming to an end.

(Q) Frightened?

(A) Uh huh. I thought it would hurt you, and it did.

(Q) What do you think it was?

(A) I don't know.

(Q) Feel bad the next day?

(A) Uh huh.

(Q) See helicopters?

(A) Yes.

(Q) When?

(A) The same time I seen the flying saucer.

(Q) Where?

(A) All over.

(Q) When the car was stopped did you see helicopters?

(A) Yes.
(Q) How many?
(A) I counted 23 of them.
(Q) What were the helicopters doing?
(A) Trying to find out what it was.
(Q) You didn't see them coming?
(A) No.
(Q) When did the helicopters show up?
(A) About the same time, we saw the light over.
(Q) See where they came from?
(A) No.
(Q) See any people in the helicopters?
(A) No. I didn't see anybody.
(Q) Helicopters always ahead or over the car? (A) Ahead.
(Q) How long were you sick?
(A) About a week or so.
(Q) Any bad dreams?
(A) Yes.
(Q) What about?
(A) Same thing we saw.
(Q) How often?
(A) Almost every night. I had one last night.
(Q) Are you still frightened by it?
(A) Sometimes.
(Q) You didn't get any bums?
(A) No.
(Q) Is there a spot on your cheek?
(A) Left cheek.

He showed a spot about the size of a quarter.

(Q) Did you tell your friends about it?
(A) Yes.
(Q) What did they say?
(A) They didn't believe me.
(Q) Does that bother you?
(A) Yes.

(Q) They still don't believe you?
(A) They believe me now.
(Q) What made them change their minds?
(A) Don't know.
(Q) What color were the flames?
(A) Yellow.
(Q) What did it sound like?
(A) There wasn't sound to it. I just heard the helicopters.
(Q) When you got back in the car, what did you do?
(A) I hid my face.
(Q) In front seat?
(A) Yes.
(Q) On seat or on the floor?
(A) I was in the seat.
(Q) How did you hide?
(A) Like this.

He covered his face with his arms.

(Q) Why?
(A) Cause I didn't want to get hurt.
(Q) What makes you think it would have hurt you?
(A) The lights and all that.
(Q) The lights were bright?
(A) Yes.
(Q) Did it hurt your eyes to look at it?
(A) Yes.
(Q) You say you were out of the car for three minutes?
(A) A few minutes.
(Q) What did you say to grandma?
(A) I said get back in the car.
(Q) Did you think something was going to happen?
(A) Uh huh.
(Q) You got pretty scared that night?
(A) Yes.
(Q) Does it bother you that you got scared?
(A) Yes.

(Q) Why?
(A) I don't know.
(Q) Don't you think everybody would have been scared?
(A) Yes!

CHAPTER 12
Sensory Data—Sounds & Heat

In addition to the visual effects of the bright light, this encounter also includes the clear recognition of a number of sounds and the feelings of extreme heat. Both of these conditions are well recorded and are important to the overall investigation.

The Radio Stopped Playing

Before the encounter they were having an enjoyable drive home, chatting about how "things used to be," with the radio playing softly in the background. When the object came down from the sky and hovered over the road ahead of the car, none of them payed any more attention to the radio or the sound of the car's motor.

Betty did clearly remember leaving the car running when she stopped and got out. She wanted to be able to get away if possible. After the object left, however, Betty was hoping to hear some news about the event, so she was concerned that the radio was not playing until she was some distance from the object.

Betty told what happened when she got back in the car, as follows:
(Q) And when it (the object) moved, that's when you felt for the keys?
(A) Yes.
(Q) And turned the car on?
(A) That's when I realized the motor had gone dead in the car, when I had to turn the key.

(Q) You put it in park when you stopped?
(A) Yeah.
(Q) But you didn't turn it off?
(A) No. I was going to leave that thing going, so I could go whenever that thing went. Or seen what it was going to do.
(Q) When it did move up, when did you realize the engine was not running?
(A) When I got back in the car, and as soon as that thing lifted up, I was going to start the car up. I put on the gas and everything and it wouldn't move.
(Q) And you had to start...?
(A) Start the car.
(Q) Did you turn the lights off or anything to start the car?
(A) No. I didn't touch anything except the car key and the gas pedal.
(Q) And the radio, you say?
(A) It was dead.
(Q) No sound was coming out of it?
(A) No, there wasn't any sound.
(Q) Did you turn it off then, or?
(A) No.
(Q) Left it on?
(A) Left it on, and then going on the highway, we thought we'd hear it. After we got on out by the stop sign to turn onto 1960, the radio started playing again. We thought, well, we could hear something about what the helicopters were doing and what this object was, maybe, before we got home. But we never did.
(Q) But the radio didn't come on until you got down to 1960?
(A) Uh huh.
(Q) Did you hear any kind of static during this time?
(A) Didn't hear one sound out of it.
(Q) Not until you got to 1960.
(A) Uh huh.
(Q) Where, just as you turned onto it or what...?
(A) No, we had to stop at a stop sign at 1960. Just as we got to that stop sign, there, the radio came back on.
(Q) Did you ever have any trouble with the radio before?
(A) No.

(Q) How old was the car at the time? When did you buy it?

(A) In December 1979. It's an '80 automobile. They had just come out with it.

(Q) So, you had it just about a year?

(A) Uh huh.

They Heard A Beeping Sound

Betty and Vickie talked about the beep-beep sound they both heard throughout the encounter. It didn't matter whether they were inside or outside the car. From their description, it appears the sound they heard had the same low-frequency amplitude modulation as was recorded in several Canadian UFO incidents a few years earlier. Betty was quite animated as she answered questions about the beeping.

(Q) One of the stories quoted you as saying you heard a beeping sound that was so shrill you thought it would break your eardrums.

(A) I still have ear problems from it. I believe I always will in my right ear.

(Q) Did you think it was going to break your eardrums at that time?

(A) Yes, I sure did.

(Q) The beeping sound itself was so shrill?

(A) Right.

(Q) Did you hear the beeping sound before you stopped the car?

(A) No, because I had the radio going. I feel that that could be the reason I didn't hear it, plus Vickie and Colby and myself was all upset and talking and everything. I....I really wasn't paying that much attention to it.

(Q) They heard the beeps also.

(A) Yeah. You couldn't help but hear it—it was unreal, really.

(Q) Would it hurt your ears when it beeped?

(A) Oh, yeah, it would hurt your ears! Yeah, it was just like—well, I've never heard anything to compare with it in my life.

(Q) Was it a long beep, or would it...?

(A) No, it'd go BEEEP!...BEEEP! BEEP! like that.

(Q) But extremely shrill, huh?

(A) Oh, yeah.

(Q) High pitched?

(A) Very high-pitched.

(Q) Did the beeping coincide in any way with the flames shooting down?

(A) No. It beeped constantly.

(Q) Constantly? Throughout the whole thing?
(A) Right.

Sound Associated With The Flames

In addition to its overall brightness, the object would periodically shoot bright flames downward toward the surface of the road. From time to time the flames were accompanied by a loud sound which added to the frightening nature of the event. After Vickie's narrative description of the event was finished, she was questioned about the sound she heard.

(Q) Was it a real strong flame?
(A) It was a real strong flame. And then when it would let up, it would go up.
(Q) But you would still see the flames when it let up?
(A) Uh-huh. And then when it would blow again, it would come back down...
(Q) You've seen on TV and movies pictures of missiles and rockets, space shots, and things like that...?
(A) That's the way it was, but when the flame came down, it wouldn't fly off.
(Q) But the flames would come down real strong?
(A) Uh huh.
(Q) Was there any smoke involved?
(A) I didn't notice any...
(Q) When the flames came down, you could hear the whooshing sound. How would you describe it?
(A) Well, you know how, have you heard a tornado? (answer - no). You've heard the roar of big winds? It was like the roar of an engine is—but you've never heard such a roar in your life.

Under cross-questioning, Betty also described the unusual nature of the sound coming from the flaming object.

(Q) What did you observe?
(A) It was a huge.... a bright silver....an aluminous thing. It was diamond-shaped, with fire coming out of the bottom. When I heard "that air thing," that's when the flames would shoot out....
(Q) Doing this periodically?
(A) Right. It sounded like air. You've heard these air brakes on these big old trucks when they go hitting them real hard? That's about what it sounded

like, but louder. I mean, it sounded like air—like a pressure.
(Q) You heard this only when the flames came out?
(A) Really, yeah.
(Q) When you didn't hear this, were there any flames?
(A) Yes, there sure were.
(Q) But they weren't as long or what?
(A) Well, they weren't as big around.

They Could Feel The Heat

As they approached the huge object hovering over the road, they could feel the heat coming from it. Although it was a chilly winter night, the car's interior warmed rapidly. That is when they got out and stood in the open doorway, watching the object. To make matters worse, Betty left the protection of the car's door and walked to the front of the car to get a better look at the craft. Before the encounter was over, the car became too hot to touch comfortably. Betty felt severe pain when she touched the fender and had to use her coat as a hot pad to re-enter the car's door. Later, as they drove away from the scene, she turned the air conditioner on to cool the car down.

Touch temperature is an objective measure, although not a quantitative measure. Guidelines for designing equipment for human operations in a warm or cold environment specify a touch temperature of 105 degrees Fahrenheit as the limit for surfaces that must be grasped by human hands. The NASA Bioastronautics Data Book, the official source of biomedical design data for engineers who are designing America's space vehicles, lists the pain threshold for any area of skin as 113 degrees. And the typical sensation for touching a surface above 95 degrees is defined as unpleasantly warm. From this we may deduce that the temperature of the car's fender, as well as the driver's door, where it was touched by Betty that night, was above 113 degrees.

Vickie said: "The light was there, the object was there, and it was just like any minute we were going to get burned up."
(Q) When did you first feel the heat?
(A) Almost instantly.
(Q) When you stopped the car, you began to feel the heat?
(A) Uh-huh.
(Q) When you opened the door, was it even hotter, or was it about the same?

(A) It was hotter...

Betty had the longest direct exposure to the heat from the object because she was outside the car for the longest period of time. It is speculated that is why her burns were more extensive than Vickie's.

Why did she stay outside so long? Betty said: "I felt safer outside because I thought at least out here I'm not going to be pinned in that car and burned because I can break and run and maybe get away from some of the heat."

(Q) When you opened the car door, did it suddenly get hotter?
(A) Oh! Did it ever! It sure did. Very definitely.
(Q) What did it feel like?
(A) Well, I've never been burned—except a little blister from cooking or something, but it wasn't that kind of heat.
(Q) Have you ever been to a bonfire? A pep rally type of bonfire? Ever been close to a fireplace?
(A) Yeah. It was hotter than that.
(Q) Hotter than any fireplace, huh?
(A) Oh yeah.
(Q) When you opened the door, a big wave of hot air?
(A) Yeah, but the car was getting hotter too.
(Q) The car — inside—you mean the car itself was feeling hot?
(A) Oh, yeah. Even the steering wheel was getting hot.
(Q) You have a plastic steering wheel, don't you?
(A) I don't know.
(Q) What kind of upholstery do you have in the car?
(A) It's leather—on the dash, it's leather.
(Q) What about the seats? Are they vinyl or?
(A) Velour.
(Q) Could you feel the heat on the velour?
(A) Yes, I could—all the way through the car.
(Q) And when the flames receded, it would cool off a bit?
A) No. It didn't really cool off all that much.

Betty said she held her hand up to shield her eyes so she could get a better look at the object. She said the light was "just blinding." She described how she walked forward to get a closer look. She went as far as the headlight but never got out of reach of the car. During the re-enactment of the event, she reached out and touched the fender of the car, but she described it as the hood of the car. The responses to questions about touching the car provide some good insight into how hot it was.

(Q) What did you touch, the fender or hood or what?

(A) The hood.

(Q) Why did you do that?

(A) Well, you know, cars don't have fenders anymore. I just went past the headlight on my car.

(Q) And you reached over. Why did you do that?

(A) Because I was blinded, and I couldn't see.

(Q) And you reached over and touched the car to make sure where you were?

(A) That's right. And when I did, I burned my hand.

(Q) And what did you do with your hand?

(A) Well, I jerked it back.

Betty went on to tell how her hand was treated with salve at the hospital in the same manner as the burns elsewhere on her body. It was months before that hand would be the same again. Then she responded to questions about how she got back inside the car.

(Q) When you got back in, you said you had to use the coat tail of your coat to close the door. Why?

(A) Because the handle was so hot when I started to touch it. The handle was so hot I couldn't stand to touch it.

(Q) You are talking about the handle.

(A) On the outside of the door.

(Q) Was the door closed?

(A) Yeah.

(Q) Is this one (handle) where you lift up?

(A) Yeah.

(Q) And you had to use your coattail to...

(A) Pull up under it to get in the car.

(Q) Did you touch the handle first?

(A) Yes, I did.

(Q) And what happened?

(A) I seen right then it was too hot for me to fool with because this (right) hand was burned. And I thought, I can't afford to have the other hand burned too, so I just got the tail end of my jacket.

(Q) With your left hand?

(A) Yes.

(Q) Did you touch the door handle with your left hand?

(A) Yes, just with the tips of my fingers.

(Q) And it was too hot?

(A) Uh-huh.

(Q) Is this a knee-length coat?

(A) No, it's waist— well, it's a little longer than waist length. They call them a car coat.

When questioned about how the air felt when they breathed it in, their response was "just hot."

CHAPTER 13
Helicopters Are Up There Too

"O h, thank God! thank God!" Betty said as the object rose and began to move away from them. Then almost immediately she blurted out: "Vickie, there's helicopters chasing that thing!"

As the object rose up, Betty and Colby could see five or six large, dark-colored helicopters flying around the object. Vickie didn't see the helicopters at first. Her eyesight isn't as good as theirs, and she was still dazzled by the brilliance.

All three had heard the distinctive chop-chop sound of the helicopter rotors a minute or two earlier, but they had thought the sound was part of the noise coming from the object. The sudden upward movement left the helicopters behind momentarily, but they caught up with the object in a second or two, staying with it as it began to move slowly away. Betty took the opportunity to start the car and make a getaway.

The object and the helicopters were off to their right but still not very far away. Betty was driving south, and they were moving at an angle to the southwest. She had thought for a second of turning around and going in the opposite direction, but she wanted to get home as fast as she could.

"I see the helicopters now," Vickie said. "I wonder what they're doing?"

"I don't know," Betty replied. "They sure are pretty weird looking themselves."

Each of the helicopters had two big rotors on it, one in front and one in back. The helicopters were much bigger than the car, but alongside the object, they appeared to be small.

"It looks to me like they're helping it or trying to hem it in," Betty said. "See how they're stacked up around it. It looks like some are trying to get on top of it." Betty stopped near a small bridge about two miles down the road and said: "Look, there's more of them (helicopters) coming over there!"

She pointed to the left, and low in the sky; they could see a steady stream of helicopters flying toward the object from the Dayton area to the east.

"I wonder why there are so many?" Vickie asked.

By now the object was a quarter of a mile to half a mile away and now it looked like an oblong ball. It glowed red like a blacksmith's anvil. They never saw any more flames coming out of it after it lifted up above the trees.

Betty drove on, watching the object and the helicopters almost as much as she was watching the road.

Vickie watched the helicopters a minute or two and said: "One thing's for sure, they know what that thing is, or they're trying to find out."

"They must know what it is, or they wouldn't be there," Betty replied. "Those people in the helicopters better watch themselves. As hot as we were back there, they've got to feel it. Some of them are even closer than we were, and if they don't watch out, they're going to get burned up."

"Maybe they don't know how hot it is around it," Vickie said.

CH-47 helicopter, the type seen at the incident.

A little further down the road, they turned west on Route FM2100, and a short distance later, Betty stopped near the entrance to a rural cemetery. They could clearly see the object and the helicopters across an open field.

At that point, they began counting the helicopters. Betty thought there were twenty-six. Vickie and Colby were counting together, and Vickie said there were twenty-one.

"There's two more over there, Grandma," said Colby. '

"Maybe I counted several twice," Betty said.

They finally agreed there were at least twenty-three helicopters flying around the object, with more flying toward it from the east.

They drove on, and about a mile beyond the cemetery, Betty turned south on the Huffman-Eastgate Road, which connects with route FM1960 about two and a half miles away. The object and the helicopters were much farther off to their right now but still were clearly visible. At the intersection of FM1960, they stopped and again counted helicopters. Again, they agreed there were at least twenty-three.

At that point, they stopped following along behind the object and the helicopter armada and turned to the east, intent on driving the 11 miles back to Dayton as rapidly as possible. The helicopters were real, and they knew it. There was no point in counting them again.

CHAPTER 14
Related Ufo Activity

A number of other UFO sightings were reported around the time of the Cash-Landrum sighting. Some of those reports are included here to illustrate the fact that a significant level of UFO activity was ongoing around the world.

RAF Woodbridge, England. December 27 and December 29, 1980

British UFO researcher Jenny Randles has written extensively on the UFO activity in England on Friday, December 26, Saturday, December 27, and Monday, December 29. She has presented substantial evidence of UFO overflights and landings.

One of the key witnesses of the December 29 incident was an American, Charles I. Halt, Lt. Colonel, United States Air Force. He was Deputy Base Commander at RAF Woodbridge. He summarized the UFO activity in an official Department of the Air Force letter dated January 13, 1981. The subject of the letter was noted to be "Unexplained Lights."

The text of the letter follows:

1. Early in the morning of 27 Dec 80 (approximately 0300L), two USAF security police patrolmen saw unusual lights outside the back gate at RAF Woodbridge. Thinking an aircraft might have crashed or been forced down, they called for permission to go outside the gate to investigate. The on-duty flight chief responded and allowed three patrolmen to proceed on foot. The individuals reported seeing a strange glowing object in the forest. The object was described as being metallic in appearance and triangular in

shape, approximately two to three meters across the base and approximately two meters high. It illuminated the entire forest with white light. The object itself had a pulsing red light on top and a bank(s) of blue lights underneath. The object was hovering or on legs. As the patrolmen approached the object, it maneuvered through the trees and disappeared. At this time, the animals on a nearby farm went into a frenzy. The object was briefly sighted approximately an hour later near the back gate.

2. The next day, three depressions 1 1/2" deep and 7" in diameter were found where the object had been sighted on the ground. The following night (29 Dec 80) the area was checked for radiation. Beta/gamma readings on 0.1 milliroentgens were recorded with peak readings in the three depressions and near the center of the triangle formed by the depressions. A nearby tree had moderate (.05-.07) readings on the side of the tree toward the depressions.

3. Later in the night, a red sun-like light was seen through the trees. It moved about and pulsed. At one point it appeared to throw off glowing particles and then broke into five separate white objects and then disappeared. Immediately thereafter, three star-like objects were noticed in the sky, two objects to the north and one to the south, all of which were about 10 degrees off the horizon. The objects moved rapidly in sharp angular movements and displayed red, green, and blue lights. The objects to the north appeared to be elliptical through an 8-12 power lens. They then turned to full circles. The objects to the north remained in the sky for hours and beamed down a stream of light from time to time. Numerous individuals, including the undersigned, witnessed the activities in paragraphs 2 and 3.

Charles I. Halt, Lt. Col, Usaf Deputy Base Commander
Echols. Kentucky. December 28, 1980

The December 30 issue of the Owensboro, Kentucky, *MESSENGER & INQUIRER*, reported UFO sightings in Ohio County, as follows:

"ECHOLS—It was no fantasy," said sheriff's deputy Frank Chinn of Ohio County, adding, "and I don't drink."

Chinn, at first reluctantly, was explaining how he sighted eight brightly colored, unidentified flying objects in Ohio County Sunday night.

Among the witnesses were another sheriff's deputy and a Kentucky State Police trooper.

In addition, the Ohio County Sheriff's office received several reports of the sightings Monday after getting 14 such calls Sunday night.

Chinn said he was at his home in Echols, about one and one-half miles south of Rockport, when six objects following an identical flight path began traveling slowly out of the southeastern sky, beginning about 5:30 p.m.

The deputy called for verification from State Police, and Trooper Don Beemer said he arrived in time to see one of the objects leaving.

Chinn said he saw two more UFOs later in the evening, at times using a pair of binoculars. Chinn said he and fellow deputy Johnny Cooper also watched one UFO through a telescope set up on a tripod at a Centertown gas station. Chinn said he saw rotating red, green, and yellow lights, which through the telescope resembled lights through the facets of an upside-down diamond.

Chinn, a former Marine, said, "I don't know any aircraft of ours that could move like that—so slow and no sound. It was definitely not a helicopter.

Beemer said that by the time he arrived, the object looked like a bright star changing colors. Beemer, a Vietnam veteran, said he was sure what it wasn't: "It was no airplane, and it wasn't a helicopter. It took forever to move across the sky."

The Ohio County Sheriff's radio dispatcher said that she contacted a UFO reporting center and the Henderson State Police post but that there were no reports of downed aircraft, search parties or off-course weather balloons."

Hartford, Kentucky. December 28, 1980

The January 8, 1981, issue of the Hartford, Kentucky *OHIO COUNTY TIMES*, reported four more silver objects that were seen on December 28. The text of that article follows:

"It is a common characteristic in a majority of parents to take lightly much of what is said by their children—especially very young children.

Bobby and Christine Williams, Hartford, learned last week that even the innocent ramblings of two six-year-old boys are worth bending an ear to.

Even so, the nature of what the youngsters were talking about would prompt the average parents to be a little less than understanding.

Jason and Jesse are the twin sons of Mr. and Mrs. Williams. What they saw on the evening of December 28—strange as it may seem—apparently was seen by several other Ohio Countians.

More fiction than fact has been attached to the sightings of unidentified flying objects (UFOs), but nobody in an official capacity has been able to disclaim their actual existence.

Jason and Jesse were not capable of placing an accurate identity on what they saw in the sky near their home but whatever it (or they) was created a level of excitement in their young imagination.

"I didn't pay too much attention," Mrs. Williams recalled. "The boys were in the front yard and came back in excited about something they saw in the sky."

It was not until after a story appeared in the Wednesday edition of an area daily newspaper did the mother attach any importance to what her young sons attempted to say. Reports of strange lights over the Olaton area came from several residents, including a Hartford police officer. Officials concerned with possible UFOs have been unable to explain the reported sightings.

That, however, does not satisfy the curiosity of young Jason and Jesse Williams. Although their stories differ slightly, the boys unmistakably believed that what they saw was not an everyday happening in the cold, winter sky.

Jesse, in a tone as authoritative as a six-year-old can muster, said he saw some "silver things" in the sky and that one of them dropped something. "It looked like a bomb or missile, but I didn't see it hit the ground," the youngster said.

Jesse said two of the flying machines went straight, and the other two went off in other directions.

"They were flying like a plane.... about 10 feet up," Jesse recalled. Jesse said the "silver things" did not make a noise and moved slowly. Jason saw "little triangle things flying through the air—two in one direction and two in another." Unlike his twin brother, Jason did not see anything drop from one of the craft. "I just watched for a minute and then had to go in," he said.

Regardless of what they saw or what it actually was, the sightings did not upset the Williams twins. "We didn't know what it was, so it didn't scare us," Jason said.

Cleveland, Texas. December 29, 1980

A maintenance engineer at Houston's Methodist Hospital was returning home to Cleveland from Houston after completing his second shift job; at the hospital. It was around midnight on December 29, 1980. Just after he crossed the bridge on Interstate 59 just south of Cleveland, he came across a UFO at treetop level. It was the length of two football fields. He got out of the car and observed it for 20 to 30 minutes. He said it was close enough to see beings inside the vehicle, and some seemed to be looking out. The craft made no sound. When it left, the object went up and away rapidly.

Near Dayton, Texas. December 29, 1980

Belle Magee lives seven miles west of Dayton. She was sitting at a table playing cards when she saw something out the window in the direction of New Caney. The time was about 9:00 p.m. She said: "It was just a light, a big old thing. I knew it was something that wasn't supposed to be there. It wasn't there that long. I said, 'look at that light!' My son turned around, and by the time he looked it was gone."

Near Huffman, Texas. December 29, 1980

Angela and Doug Stanley of Dayton were driving back to Dayton from New Caney. They had just completed driving down the long straight stretch of FM1485; the same area where the Cash-Landrum encounter occurred. As they turned on to FM2100, they saw a very bright light in the sky. They said it was very bright and looked 'like a car in the sky.' They had no reason to stop and search the area.

Dayton, Texas. December 29, 1980

Jerry McDonald was working in his yard between 7 and 9 p.m. when he saw an object as large as the Goodyear blimp overhead. He distinctly remembers it was Monday night because that is the night, he took the trash out for pick up. He said he was in the back of his trailer house, trying to fix a leaking water hose, when he heard the object approaching. It had a low rumble to it. His description follows:

"It was kind of a triangular shape. As it came over, it was about 150 to 200 feet (up), and it was going maybe three to five miles per hour. I could see these funny-looking lights. It had white and blue lights on the corners and a bright red light in the middle. The back had what looked like two jet propulsion type engines with the fiery torch looking color, like an acetylene torch."

As the object passed directly overhead, he could plainly see it was triangular in shape. It continued on over the vacant football field and out of sight to the west, in the direction of Huffman.

By Wednesday, Jerry had flu-like symptoms, with a 103-degree temperature and bad headache. He was sick off and on for the next six weeks. On February 14, he was hospitalized because of 'an air pocket' in his lung. VISIT investigator Dave Kissinger was so concerned after his meeting with McDonald on March 23, that he thought the State of Texas Health Department should be notified of the potential hazard. A follow-up interview several weeks later by Bob Pratt found that McDonald had recovered from his problems.

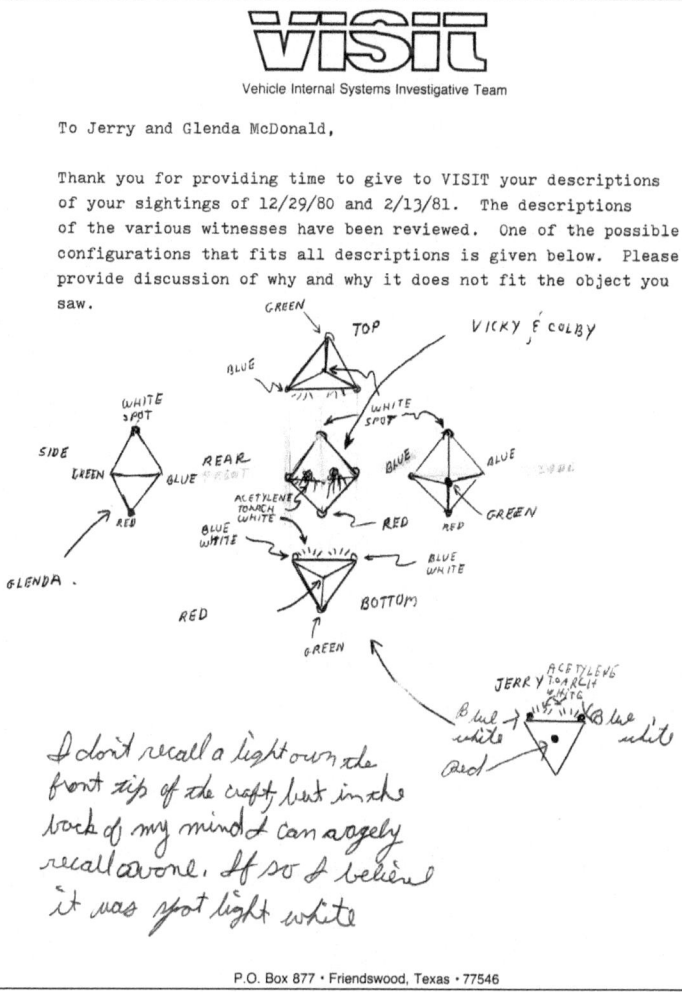

Project VISIT report of the sighting by the McDonalds on the night of the incident.

CHAPTER 15
Baseline Health Status

Following their exposure to the object that night on route FM1485, Betty, Vickie, and Colby experienced numerous and lasting medical problems. Those problems were monitored and documented as they happened, and medical records were obtained and held for future reference.

With the exception of Betty's earlier heart operation, the trio claimed to be in good health before the event. They openly shared past records, doctor's names and addresses, and contact information for people who could verify their claims. Some of the statements made about their pre-incident health conditions are included in this chapter.

This material represents a cross-section of the database that has been amassed, showing that Betty Cash, Vickie Landrum, and Colby Landrum were in reasonably good health before the incident. Following the incident, their poor state of health is a matter of public record, backed by medical records and photographs.

STATEMENTS ABOUT BETTY CASH
Statement By Sharon A. Jones:

"I, Sharon A. Jones, a niece of Betty Cash, am writing this letter concerning her health prior to the military incident on December 29, 1980. I wish to verify that excluding a previous open-heart surgery, my aunt was in good physical health. However, following the incident, I have witnessed a steady decline in Betty's health."

Statement By Sandy Lebeau:

"This letter is regarding the health of my aunt (Betty Cash) before and after her exposure to radiation.

Since my aunt's exposure to radiation, I have watched as the poorness of her health has at times gotten the best of her.

To think that she has gone through being burned, had a loss of hair, has had a mastectomy, not to mention the emotional upset since the sighting infuriates me to think that our own government is taking no course in helping her.

My aunt and I have always been close. I watched her go through heart surgery, in which she came out smelling like a rose.

She has always had an abundance of energy. She's always had the ability of making people laugh and enjoy themselves just by being near her.

The change has taken its toll on her and her loved ones. She hasn't half the energy she used to have and when undergoing chemotherapy, she's often too sick to get out of bed.

I have to say this incident has really shaken my belief in our government and its military."

Statement By Lois Green:

"I, Lois Green, am the sister to Betty Cash. We have been very close all of our lives. It really hurts me to see her in the physical health she is in now compared to the health she has always had. She had a heart attack, and she was able to bounce back and be herself again, but I have never seen anyone suffer as much as she has since December 29, 1980, when she was exposed to the radiation from the object flying in the sky.

I cannot understand any human being thinking this is not what has caused her health to gradually get worse. I know for I was the one to care for her and bought her a wig and also saw the blisters form all over her and hear the doctors telling us how thin her skin is from the radiation and staying with her night and day when she has operations and other complications that has never happened before.

I know this is ugly of me, but I pray that if anyone does not believe what this did to her that they will never have to witness someone they love suffering like she has, but that person himself will have to live a life that will be as painful and every way they turn someone will tell them that they are not sick and nothing is really wrong. Betty enjoyed life so much prior to this incident, and now she is in pain all the time plus being sick at stomach so much."

Statement By Jesse L. Collins:

"I, Jesse L. Collins, Betty Cash's brother, visited her in the hospital the day after she was admitted, and here is what I saw: her hair was falling out—her skin on her head was blistered and (her) neck and arms had blister-like scars, and I did not want to get too close to her."

Statement By Mrs. Mary F. Bierbaum:

"I visited Mrs. Betty Cash a few days after she was admitted into Parkway Hospital. Her condition was as follows: eyes set back and dark rings under her eyes; hair falling out in masses; and blister-like sores covering her body. I did not visit long and did not get close to her due to her condition."

Statement By Joyce Ferguson: *(Beautician – when questioned by Bob Pratt)*

(Q) How long have you been taking care of Betty's hair?

(A) I guess about three years, off and on.

(Q) What was her hair like before that (the incident)?

(A) It was perfect and thick. She always had the thickest hair of anybody I'd ever seen. I could almost do two people before I could do her. And it was just perfect and not a spot on it anywhere.

(Q) OK, you said it always took you about twice as long to do.

(A) Yes, uh-huh. It took me about twice as long to do her as anybody because it was so thick. She had three hairs where everybody else had one (laughs).

(Q) Do you recall the last time you worked on Betty's hair before this happened?

(A) Uh, it musta been a week or two ...

(Q) In the two or three years that you treated Betty; did she have any trouble at all with her hair?"

(A) No, sir. I have never seen any kind of thin spot or any kind of problem at all. A lot of times I wished some of it would come out (laughs) before I got through with her. Bless her heart. But she was just fortunate enough to have a good head of hair.

Statement By Wilma Emert: *(Stayed with Betty the night she was hurt)*

"She was hurting all over. Her face was real red, and she complained about her neck, and she asked me to feel and see if there were any knots on her neck. I didn't, but her son Toby did. She got bad real fast. Her left eye was swollen shut.

She told me about the UFO that night, but I didn't even listen to her. I had already had a few drinks. But my kids heard her (niece Leslie Marie, 10 and son Darrin, 15)."

Statement By Mickey Joyce Foster: *(Betty's daughter)*

"I walked through the door (at Parkway Hospital), and I turned around and walked out because I thought I was in the wrong room. Her face, the doctor told me, was the size of two men's faces put together. Her eyes were swelled closed. She didn't even know I was there until I spoke up, and I'd been sitting there about ten minutes just trying to get my head together."
"You mean you were in a state of shock from seeing her or what?"

"Yeah, I was. You know? You just don't walk into a room and expect to see your mother, and it looks like someone else laying there. This was on January 4, 1981. Dr. Shenoy couldn't explain what was wrong with her.

That's another thing, too, like on her forehead. It looked like she had been under a sun lamp. And when I rubbed her forehead, water just squirted out. It looked like a blister that had been popped. But it was all over her face. And it was all in her hair."

Statement By Dr. V.b. Shenoy:

"Miss Cash was admitted to the hospital on January 2, 1981. Her main problem at that time was a severe headache, a swelling of the face, a swelling of the eyelids, and a swelling of the skull on the top of the head. One of the small glands on the back of the neck was swollen."

"She stayed in the hospital approximately 12 days. The headache was getting better. The swelling of the face, eyelid, and skull was getting better, so she went home on the 14th of January. A few days after going home, she felt that the headache is coming back again, the swelling of the skull is coming back again, and she noticed, uh, her hair is getting thinner and thinner, and four or five days afterward she noticed she is losing hair in patches. So, she came back to the hospital. And when I saw her, I was surprised that the hair was gone on the head in patches, four or five big patches on the skull, measured at least three or four inches in diameter."

Dr. Shenoy replied as follows when questioned whether or not the condition could be related to her earlier heart problem:

"Whatever symptoms she had had nothing to do with the underlying heart problem. She had bypass surgery in Alabama sometime in 1977. She did not have any of the problems like high blood pressure, diabetes, or the tension headache, and she responded well to the pills we gave her for the heart. So, she sees us or one of our associates every six months, and the problem is all gone. The swelling of the face and the intense headache, this will be the first time I came across this with Miss Cash. She did not have the symptoms in the past."

Statement By Dr. K.b. Fung: *(At the time of her initial hospitalization)*

This statement is shown here to establish a baseline condition with respect to the breast cancer that developed in 1983.

"Opinion - There is no remarkable change seen in the xeromammography examination bilaterally." In other words, it was not a pre-existing condition.

STATEMENTS ABOUT VICKIE AND COLBY LANDRUM
Statement By Louise Landrum: *(daughter-in-law)*

"My name is Louise Landrum, daughter-in-law of Vickie Landrum. I have been in the family for 23 years and, to the best of my knowledge, have known of no kind of illness for Vickie, outside of a minor cold, a few times. She was in the hospital only once that I can recall for the flu."

"She was putting in from 16 to 17 hours at the cafe where she worked before the incident happened. All you have to do is to check with some of the customers, and they can vouch for the amount of time she was capable of putting in. As far as I know, she was in good health until after the incident."

Statement By Jayne Magee:

"This letter is to verify that my mother is telling the truth. I am stating fact that her health and Colby's were indeed in good condition until "after" the incident. She used to be able to carry on and work like a 20-year-old until then, and now she has trouble getting out of bed at times. Whoever is responsible should take care of her and Colby and Betty, for they have lived through hell and will probably do so until they die because of the incident."

Statement By Jean Burnett:

"I am writing to let you know that my mother would not lie about something like that. She would not burn Colby and herself. Or she would not pull out her hair. All anyone has to do is talk to anyone in Dayton or Liberty County that knows her and find out that she is a very honest person. They can check with the doctor about her health before and after this happened. My mother was in good health. I lived in Abilene at the time this happened and had to move back to take care of her and Colby. Her eyes were so bad she could not do anything. Her eyes are not any better, and her health is bad. There are days that she can't get out of bed."

Statement By Martha Thompson:

"Before Vickie Landrum's encounter in December of '80, she worked for me, and her health was very good. Her eyesight since then has gotten very poor. We used to go fishing and play bingo all the time, but now she can't see to bait a hook, and the sun causes big blisters on her when exposed. She can't see well enough anymore to even hardly read a bingo card. She gets sick to her stomach a lot and has lost a large amount of hair also."

"Colby, her grandson has had a lot of dental problems since their encounter and there for a while had trouble with his bowels and sleeping properly."

"Knowing both of them as I do I know what a great mental and physical effect this has made on them both. Vickie can no longer work because of bad eyesight and feeling bad all the time and Colby still lives his terror in his dreams. I swear all the above is true to my knowledge."

Statement By Jim Mcmanus:

"I have known Mrs. Landrum for over ten years. Prior to December 29, 1980, she was in good health. She has always been able to work and be on the go without any problems. But since the night on December 29, 1980, she has been sick and at times can hardly go at all. Whatever happened that night certainly has caused her health to deteriorate greatly."

Statement By Laura Mae Landrum:

"David and I were not in the Dayton area when his mother, Mrs. Cash, and Colby were involved in the disturbing happening of December 1980, but I would like the whole world to know that we believe in them and that theirs is the unvarnished truth."

"We moved back to Dayton in June of 81, and other than one visit, it was the first time we had seen the family in almost a year. The change in Vickie Landrum was astonishing. Since I first met the family in 1968, she had always worked and tended to their family, making sure they had all they needed. She told me once that growing up in the depression made her want to make sure her children never went hungry or wanted for anything. Anyway, I digress; the change I mean to note was that not only was she not working because of her eyes, the unsightly sores, and her loss of hair, but she has become almost a hermit. She very seldom goes anywhere and when she does is met by skepticism and questions. "

"I have seen her go to bed at night, and the next morning, when she got up, her left eye would be swollen and running and so inflamed it appeared she had sunburn. For days at a time, she has headaches, and these blisters just seem to pop out as if they have been in hiding."

"David plays Softball, and Vickie has grandchildren involved in sports. In the very late evening and night, she can attend (games), but if she attempts to go to one during the day, the sun and heat will make her sick sometimes for a week."

Statement By Dayton. Texas Police Chief Tommy Waring:

"I don't know Betty Cash all that good, but I've been knowing the other one (Vickie Landrum) for a number of years, and she's never lied to me that I know of, and I've had several of her kids in Little League ball teams over the years, and as far as I know she's a truthful lady."

When asked if she would make up stories like this, he said: "I don't think so."

Statement By Vickie Landrum:

"Anyone could have checked to see my, and Colby's health was good until December 29, 1980, with a Dr. E.R. Richter—dead now. My and Colby's record has to yet be there. He treated me from 1962 until his death. When

I could not get him, I used a Reginald Wilson at 258-2624. They doctored me for colds when I had one. I was operated on in 1960. Dr. Richter did that too. Whenever I need one, they was it. Only when I hurt my leg, I used a Dr. John T. Pegues, 1409 N. Travis in Liberty. My health was real good. And I got some reading glasses from T.S.O. in Baytown. I'm sure if I need to, I could get my records. I'm not trying to hoax anyone. After we was hurt on December 29, 1980, we had nothing but hurt and misery. If the object had not been there, I would yet be fine. I sure did not burn myself or my child, hurt our eyes, or pull our hair out!"

CHAPTER 16
Betty's Medical Condition Is Defined

Betty entered Parkway Hospital in Houston on January 2, 1981 and was discharged on January 19. She re-entered on January 25 and was again discharged on February 9. This began the succession of hospital stays for Betty that have continued to the present. (Editor's Note: Betty Cash passed away on December 29, 1998)

Examination By Dr. V.b. Shenqy

Betty appeared to be a burn patient when she was admitted to Parkway Hospital on January 2 by Dr. V.B. Shenoy. Her stated complaint included swelling of the eyes, scalp, and face, along with a terrible headache. These conditions still persisted when she re-entered the hospital on January 25. However, Dr. Shenoy noted she also had a marked alopecia, or hair loss, greater on the right side than on the left. She also had swelling of the eyelids. Dr. Shenoy had specifically noted that Betty had little if any, hair loss when she entered the hospital the first time. His report includes the following information:

PHYSICAL EXAMINATION: This is a well-built, well-nourished white female in no acute distress.

VITAL SIGNS: Blood pressure 120/80. Respirations 20 per minute. Temperature normal. Heart rate 80 per minute and regular. HEAD: Normocephalic. The patient has severe swelling of the scalp associated with crusting and erythema all over the scalp. The patient also has swelling and erythema over the eyelids. There is no evidence of anemia or jaundice. Pupils are normal in size and shape, reacts well to light and accommodation. Oral hygiene is good. NECK: Supple. No cervical lymphadenopathy. Carotid upstroke is good. No bruit heard over the carotid artery or subclavian artery. Trachea in the midline. Trachea not enlarged. Jugular venous pressure flat at forty-five degrees.

CHEST: Symmetrical on both sides. Moves well with respirations. Vesicular breath sounds heard all over the lung fields. No wheezing, crepitations or pleural friction rub heard. CARDIOVASCULAR SYSTEM: Point of maximum intensity not felt. First and second heart sounds are normal in character. There is no murmur, gallop, or pericardial friction rub heard. ABDOMEN: Soft. No hepatosplenomegaly or ascites. No bruit heard over the aortofemoral vessels.

EXTREMITIES: No pedal edema or calf muscle tenderness. Proximal and distal arterial pulsations are well heard.

Dr. Shenoy's report on January 25 provided the following laboratory results: Laboratory data revealed SMA-12 showing mild elevation of alkaline phosphatase, LDH, and serum triglycerides. ANA was negative. Vitamin B-12 serum level was normal. Quantitative analysis of heavy metals from the hair was insufficient. CBC revealed mild normocytic, normochromic anemia. Urinalysis was normal. RPR was nonreactive. Latex RA test and ANA were negative. Fasting two-hour postprandial blood sugar was normal. Glycohemoglobin, however, was mildly elevated. T-3, T-4, and T-7 index was normal. Chest x-ray revealed minimal atelectasis in the lingular segment of the upper lobe of the left lung. Xeromammogram was negative. Electroencephalogram was essentially normal. Skull biopsy was consistent with alopecia areata. TEST: LEAD, ARSENIC, AND MERCURY (HAIR): Quantity not sufficient to process in test conducted on February 9. VITAMIN B-12: 1036H (reference values: 200-900) HEMOLYSIS (January 26): Albumin: 3.4 g/dl. Alk. Phosphatase: 120 U/L. BUN: 9 mg/dl. Calcium: 9.0 mg/dl. Cholesterol: 257 mg/dl. Glucose: 187 mg/dl. LDH: 309 U/L. SGOT: 17 U/L. Phosohorus: 3.2 mg/dl. Total Protein: 7.1 g/dl. A/G: 0.9. Globulin: 3.7. Total Bilirubin: 0.3 mg/dl. Uric acid: 5.7 mg/dl. Sodium: 135 mEq/L. Potassium: 4.9 mEq/L. Chloride: 97 mEq/L. CO2: 25mEq/L. Iron: 57 mg/dl. CPK: 10 U/L. Creatinine: 0.8 mg/dl. Triglyceride: 256 mg/dl. BUN/Creatinine: 11.3. Balance: 17.9.

URINALYSIS (voided on January 4): Specific Gravity: 1.008. Blood: 2+. pH: 6.0. RBC: 4-8/HPF. WBC: 3-5/HPF. Bacteria: few.

BLOOD TEST (taken on January 2): WBC: 8.1. RBC: 3.1. HGB gr: 12.9, HCT gm: 27-. MCV: 91. MCH: 31.8. MCHC %: 35.1. DIFFERENTIAL: Poly: 81, Stab: 8, Lymph: 9, and Mono: 2. PLATELET: Appear normal.

Drugs administered for the purpose of treating her injuries and relieving her pain included the following: Keflin, IV DSW 50ML, Lanoxin tab, Persantine, Inderal, Dalmane, Tylenol, Benadryl, Solu-medrol, Prostaphlin, Ampicillin, IV N.S. 0.9%, Talwin, Betadine solution, Domeboro eff. tab, Fedsol SP, Premarin, Sinequan, Motrin, Amoxicillin, Zomax, and Dalmane.

Evaluation By Dr. K. Kumar

Dr. K. Kumar saw Betty for an evaluation of the headache. She said Betty was admitted "with cellulitis of the face, especially around the eyes and had a severe headache." Neither the EEG nor the CAT scan revealed any abnormalities.

On January 29, Dr. Kumar said: "The patient had been doing fairly well until the sudden onset of the cellulitis. No definite etiology of the cellulitis was established during the last admission; but she was treated with antibiotics and steroids and did very well. The patient was discharged, recovered well from the cellulitis, and with the headache much better, and says that as soon as she went home, she started having diarrhea and soon after that, her headaches started getting much worse, and after some time, her hair started falling off. At this time, the patient is admitted for evaluation of alopecia, diarrhea, and headaches. At the time of being seen, the diarrhea has more or less been controlled. The headaches have gotten increasingly more severe, and the patient had two injections last night."

Dr. Kumar's report goes on to say: "On examination, a 51-year-old white female looks much younger than stated age. On general examination, alopecia of the scalp is present with two large areas of complete hair loss on either side of the head in the parietotemporal region. The cellulitis of the face has markedly improved at this time. Lymphadenopathy is present in the retro auricular and posterior cervical lymph nodes. Vital signs are stable."

A summary of Dr. Kumar's medical examination is as follows:

CARDIOVASCULAR: A scar is seen in the midline in the chest. Otherwise, heart sounds are well heard, and no adventitious sounds are heard. No murmurs are present, no bruits heard over the carotids.

CENTRAL NERVOUS SYSTEM: Cranial nerves - patient is alert and oriented. Seems to be very rational and neither very anxious nor depressed. Cranial nerves - pupils are equally reactive to light. Eye movements are full with no nystagmus being noted. Visual fields are full in all four quadrants. Funduscopic examination shows sharp disc margins and normal vasculature. The face is symmetrical, the tongue is in the midline, the palate moves up equally well on both sides, no dysphagia is present.

NECK: Supple at this time.

MOTOR SYSTEM: Power is 5/5 in all four limbs. Reflexes are bilaterally symmetrical. Plantars are down going. SENSORY: Patient says that pinprick and vibration are mildly diminished in the left foot, but this seems to be equivocal. In the upper extremities pin and vibration are symmetrical.

CEREBELLAR: Patient has no ataxia, can do a good tandem walk, is able to stand on either lower extremity. No nystagmus is seen. Finger to nose test is normal.

Examination By Dr. Joseph Darsey

Dr. Darsey conducted an eye examination on January 26, with the following results:

VISUAL ACUITY EXAMINATION: In the distance, each eye is 20/30 plus. The left eye is 20/30 plus. The skin of the lids shows a flat erythematous area under the right brow. There is a dry, scaly eruption on the forehead and lid skin. Extraocular motility is full. The conjunctiva is normal bilaterally. The corneas are clear bilaterally. Anterior chambers are deep and clear bilaterally. Ocular tension is 10 mm right eye, 11 mm left eye. The pupils are 4 mm and react 2+ out of 4 with direct response, and the consensual and accommodative reflexes are intact. There is no Marcus Gunn pupillary phenomenon. There is an iris atrophy. The pupils were dilated with Mydriacyl and Neosynephrine for intraocular exam. There is a minus a quarter circle, myoptic refractive error in each eye, but the visual acuity in the distance remains 20/30. The lenses are clear bilaterally, except for an occasional punctate white opacity in each lens. The inferior zonular attachments are symmetrically visible OU. There is no posterior subcapsular cataract or equatorial lens opacity. The disc, macula, and vessels appear normal in each eye, with zanthochromic macula with light reflex.

Evaluation By Dr. Solomon Brickman

Dr. Brickman made the following observations on January 5: The examination revealed crusted areas of her scalp and marked swelling of her forehead and edema and crusting of her right eyelid, and edema of both eyelids, and cheeks. There was pain palpated in her scalp. The submandibular area was puffy and tender to touch, as was the neck area. No weakness of handgrip, or her push was noted. There was no pain on straightening her legs from a flexed position. A clinical diagnosis of cellulitis with secondary edema of the scalp and face is made.

Later, on January 28, Dr. Brickman provided the following evaluation of the alopecia: There is no family history of alopecia, vitiligo, or thyroid problems. The examination revealed round spots of alopecia on the scalp. Within those areas there were areas of black hair regrowth. Mild depression of some areas of the scalp were noted and occasionally scaly areas which would probably correspond to the areas of dermatitis several weeks ago. Occasional hairs in the scalp are noted which are wide distally and markedly narrowed as they approach the scalp. Tender lymph node on the left side is noted. No major scalp pain is present. Touching the scalp did not elicit the tenderness that she experiences in her head, nor her headaches. Clinical diagnosis is alopecia areata.

Evaluation By Dr. K. B. Fung

The results of Dr. Fung's evaluation are as follows:

CT SCAN OF THE HEAD: Films were taken with and without infusion of contrast medium. The visualized ventricles are unremarkable. The pineal body and the choroid plexus are visualized with calcification. There is no definite evidence of localized increased or decreased density seen in the brain. The vein of Galen is visualized on the post-contrast film is unremarkable. The middle cerebral arteries are fairly well visualized and are unremarkable. Opinion: there is no remarkable change seen on CT scan of the head.

CERVICAL SPINE: The visualized bony structures are unremarkable. There is no fracture or dislocation seen in the cervical spine. Some prominence of the left transverse process of C7 is identified.

CHEST FILM, PA, AND LATERAL VIEWS: There is minimal radio-opacity seen in the left Lung Field, and it is not certain if this is indicating some minimal inflammatory changes. Otherwise, the size of the cardiac silhouette is borderline.

PARANASAL SINUSES: The paranasal sinuses are well developed and aerated. Otherwise, there is no remarkable change seen in the paranasal sinuses.

XEROMAMMOGRAPHY BILATERALLY: The left breast is slightly larger than the right side. Otherwise, there is no definite evidence of occupying mass or abnormal calcification seen in both breasts. Opinion: There is no remarkable change seen in the xeromammography examination bilaterally.

Evaluation By Dr. Joe Haden

Dr. Haden provided the following information of the January 29 skin sample laboratory analysis:

GROSS: The specimen, unlabeled, designated skin biopsy consists of a core of tan skin 0.4 cm. in diameter and 0.7 cm. in depth. Bisected and submitted entirely.

MICROSCOPIC: Slides have sections through the bisected halves of the above segments demonstrating two segments of skin. The epidermis is a well-differentiated stratified squamous epithelium. In one focus adjacent to a hair follicle opening the dermis shows the dissolution of intercellular bridges. One mature hair follicle with relatively unremarkable contained hair shaft is present, but there are fairly numerous superficially placed immature hair follicle structures some of which are surrounded by loosely arranged lymphocytic infiltrate—no evidence of malignant neoplasia.

PATHOLOGIC DIAGNOSIS: Skin biopsy, histologic features consistent with alopecia areata.

Evaluation By Dr. Taghi Shafie

Dr. Shafie conducted the EEG test on January 29, with the following results:

REPORT: The background activity consists of 9-10 cycles per second symmetrical and synchronous with medium voltage activity appearing on both hemispheres. No focal or generalized abnormality is seen during this recording.

HYPERVENTILATION & PHOTIC STIMULATION: Did not produce any abnormal changes. CLINICAL INTERPRETATION: A essentially normal EEG.

CHAPTER 17
Betty Is Hospitalized Repeatedly

Betty's wonderful mother, Pauline Collins, came to Houston to care for Betty. And as soon as Betty was able to travel, they went back to her mother's home in Birmingham, Alabama. Unfortunately, her medical problems have persisted over the years, and she has been hospitalized several dozen times as a result. Her visits to her doctor's office are too numerous to count. It is not necessary to discuss all of those events, but the following summary shows the extent of her medical problems, as documented in hospital records, for the year or so following the UFO encounter. It also clearly shows the progressive effects of the radiation exposure from her close encounter. Records of all of Betty's treatments are on file.

Betty was hospitalized from May 27 through June 1, 1981, at the Lloyd Nolan Hospital and Clinic in Fairfield, Alabama, under the care of Dr. Whittaker. Dr. Whittaker said his examination was basically confined to the skin, with the following results:

"The patient has numerous nevi, telangiectasia, seborrheic keratoses, etc. and over areas of her body, she has a splotchy, erythematous, macular rash which is confined primarily to the lower trunk, extremities and primarily to the high lower extremities including the anterior and posterior thighs as well as the buttock area. I could see no obvious blister at this time, although the patient pointed out to me a couple of areas where blisters originally occurred and apparently have opened and drained.

Although she has some tenderness from the splotchy areas, she does not describe any itching. She has an intense pain when she takes a hot bath and states she has to take cold baths. The patient also states that she has had chronic bowel problems ever since the UFO encounter."

"This patient has a ring-like, appearance to the areas of erythema, almost like a tinia or ringworm type lesion over the back; however, the lesions over the lower extremities, and particularly over the thighs are more compatible with irregular macular erythematous scaly areas. She also states that before the exposure or encounter (described in her file), that she had a fair amount of hair on her legs. Now her legs are quite hairless. She also has numerous telangiectasia over the exposed portions of her hands, arms and what is interesting is that the area on her fourth finger where her rings were seems to be a protected area. The skin is whiter; she has a few fine hairs. On feeling this patient's re growth of hair is quite fine and silky, again similar to that seen following growth after loss from radiation."

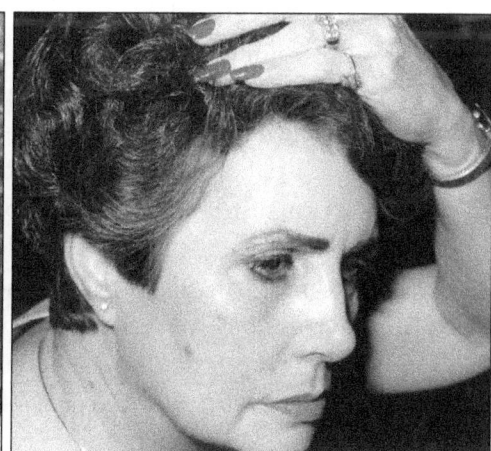

Betty's hair re-growth.

Although Betty had always been a healthy and vibrant individual, she had little resistance to disease following the UFO encounter. She was again hospitalized at Lloyd Nolan from September 29 through October 4, 1981, because of chest pains. Diagnosis revealed she had bronchitis, and she was only released from the hospital because of a death in the family. She was still cancer-free at this time, but other changes in her systems were becoming apparent as noted below.

PHYSICAL EXAMINATION: Temperature 96.4, pulse 56, respirations 24, blood pressure 128/70. Neck was normocephalic without any abrasions. Pupils equal round and reactive to light. Extra ocular movements intact. Fundi benign. Throat without erythema. There were no masses or thyromegaly. Neck was supple. Chest revealed bilateral rales. CV-regular rate and rhythm without murmur. There was no JVD. Peripheral pulses were all good. Abdomen was soft, non-tender. There were no masses. There were normal bowel sounds.

LABORATORY: CBC revealed white count of 5,500 with 60 segs, 39 lymphs, 1 eosin. Hemoglobin 13.1, hematocrit 38.1, blood gases on admission were pH 7.519, PCO2 25.5, PO2 112.5. Routine urinalysis was negative. CPK was 49, SCOT 22. Follow CPK was 24 with CPK MB band O. On admission the SMA/6 revealed a potassium 4.0, sodium 130, CO2 31, Chloride 93, BUN 9, glucose 188. Chest film on admission, with a portable chest film showed no definite pulmonary infiltrate. OCG revealed normal oral cholecystogram. Upper GI series was normal. Follow up chest films revealed the lungs clear except for mild prominence of lung markings in the left retrocardiac area. This did not appear to represent definite infiltrate and is probably not significant. Chest was otherwise unremarkable.

NUCLEAR MEDICINE REPORT: The liver is at the upper limits of normal in size measuring just greater than 25cms at the mid clavicular line. One does note a rather prominent left lobe in the mid epigastrium. There is a fairly uniform distribution of the isotope throughout the organ, arid no focal filling defects are identified. The spleen is also at the upper limits of normal in size measuring about 1 Icms, and again no focal filling defects are seen. There is a normal liver/spleen ratio in the uptake of the isotope.

Betty's condition had not improved since leaving the hospital; therefore, she was forced to return to Lloyd Nolan Hospital for another ten days starting on October 17, 1981. Dr. Whittaker was again her attending physician.

ADMITTING STATEMENT: The patient was a 52-year-old white female with a recent discharge from the hospital for presumed bronchitis. The patient was discharged at this time but before complete resolution because of illness in the family. The patient now returns without much improvement. The patient has been on antibiotics for two weeks. At this admission, the patient states no improvement, with her primary complaints being weakness, decreased appetite, diarrhea, head cold, and cough productive of yellowish sputum. She denies any fever, hemoptysis, or vomiting. The patient had a coronary artery bypass graft five years ago and has also had exposure to a high dose of radiation in December of 1980, with subsequent radiation sickness. Medications given at admission are Lanoxin, Lasix, Persantine, Inderal, Isordil, Nitroglycerin, and Premarin.

PHYSICAL EXAMINATION: On admission revealed a 52-year-old

white female, that was uncomfortable with a non-productive cough and obvious respiratory congestion. The patient looked dry at this time. Vital signs: BP 120/68, pulse 64, respirations 20, temperature 97, HEENT exam - pupils equal, round and reactive to light and accommodation. EOMs intact. There were no masses, or adenopathy. The patient did have rhinitis and sinus congestion. There were no JVD noted. Chest exam revealed coarse breath sounds. There was no wheezing or rales present. CV - normal sinus rhythm with an S4 at the apex. There was a well healed midsternal scar. There were no murmurs, rubs, or clicks. Abdomen was soft, non-tender with normal bowel sounds. No masses. Extremities - no edema, clubbing or cyanosis. Skin and mucous membranes were dry.

LABORATORY DATA: Laboratory data during the stay was essentially unremarkable. SMA 18 revealed a Trig, of 258, with the remainder within normal limits. The CBC revealed a white count of 8,800 with 66% Segs., 26% Lymph, 5% Monos and EOS 2%, 1% BASO with HGB of 12.2 and HCT of 35.3. Routine urinalysis was negative. RPR was negative. AFB of sputum was negative. She was discharged as having chronic bronchitis and chronic obstructive pulmonary disease.

The effects of the radiation exposure were becoming more pronounced as she was admitted to Lloyd Nolan Hospital again on November 26, 1981. This time she stayed until December 7.

ADMITTING CONDITIONS: Atypical chest pain. This patient is a 52-year-old white female who had onset of a left arm and chest pain on the day of admission, lasting two hours with accompanying diaphoresis. No relief from Nitroglycerine. The patient recently has been feeling fine, no cough, no sputum production, SOB or congestive symptoms. The patient also had palpations and soreness of the left costal area.

PHYSICAL EXAMINATION: The exam showed an alert, cooperative white female with BP 120/70, pulse 60, respirations 16, afebrile. Chest - few rales on both sides. CV - no gallops or rubs. Grade 2 systolic murmur, in the left lower sternal border, radiating to the apex. Aortic and pulmonic area - no JVD at 30 degrees. Extremities - no edema. EKG shows sinus rhythm, old anteroseptal MI. Chest X-ray normal.

LABORATORY ANALYSIS: Cardiac enzymes negative. The patient was admitted to CCU and treated with MI protocol. EKG showed no change from previous one, but showed an old anteroseptal MI. After 2-3 days in CCU she had a negative EKG. She had a workup for GI disease. OCG, upper

GI were negative. Cervical spine films normal. CBC was remarkable for HGB of 11.1. SMA 6/12 essentially unremarkable. Also had a CV Doppler carotid artery study that was unremarkable. Derm, consult - patient had radiation exposure as well as actenic exposure. Impression was radiation dermatitis. She had biopsy of skin and report was pending at time of discharge. The patient did well, but did have some intermittent chest pain, non-cardiac in origin. She had some numbness in the left arm. She developed a small pedal edema.

Dr. Robert Dudley, a pathologist, provided the following biopsy report:

Specimen No. 1: 0.7 x 0.5 ems. diagnosed as seborrheic keratosis. Specimen No. 2: 1.5 x 1 cm. diagnosed as seborrheic keratosis. Specimen No. 3: 0.3 cm' dia., diagnosed as hyperkeratotic epithelial hyperplasis. rhd. The samples were taken from the left dorsum hand, left mid-back, and right palm respectively.

Although she is not sure why it happened, Betty experienced the first of several falls. The first incident was on February 13, 1982, when she fell in the bathtub. She was taken to Lloyd Nolan Hospital, where it was determined that her elbow was injured and was placed in a splint.

On March 30, 1982, Betty returned to Lloyd Nolan Hospital for clinical services related to her skin condition. The dermatology consult had the same results as before.

Throughout the month of March, Betty had arm and chest pains that would last from minutes to a half-hour or longer. Therefore, on April 2, 1982, she went to Lloyd Nolan Hospital for a stress test. She felt tired and weak after the pain subsided. The stress test was terminated after 12 minutes and 30 seconds because of fatigue. She achieved a heart rate of 134/minute and had chest pain at 7 minutes into the test. Interpretation of the test results: 1) Positive submax stress test, and 2) Poor exercise tolerance.

Because her problems were continuing to mount, Dr. Luis Pineda conducted a bone marrow test on Betty on July 14, 1982, with the following results:

PERIPHERAL BLOOD: Erythrocytes show mild degree of anisocytqsis. There is a tendency toward microcytosis. Few target cells are seen. Leukocytes and platelets are normal in number and morphology. Approximately 10% of the neutrophils are polysegmented showing more than 4 lobules.

BONE MARROW ASPIRATION: Marrow particles are present.

Megakaryocytes are present in normal numbers. Myeloid to erythroid ration is 3:1. Both myeloid and erythroid series shows normal process of maturation. I do not see any infiltration by abnormal cells.

BONE MARROW IRON STAINING: There is marked decrease of stainable iron in the bone marrow.

BONE MARROW BIOPSY: Cellularity is variable arranging form areas of few to 60%. Areas of bone marrow necrosis are identified. Megakaryocytes are seen in normal number. Bone marrow architecture is preserved. I do not see any evidence of fibrosis, granuloma formation or malignant infiltration.

FINAL DIAGNOSIS: Abnormal non-diagnostic bone marrow - a) Bone marrow necrosis, b) Decreased iron storage.

In the months and years that followed, Betty continued to be hospitalized several times each year due to a continuation of the aforementioned problems.

CHAPTER 18
Dr. Shenoy Speaks Out

Betty Cash and I were participants in a press interview with Dr. Shenoy on June 16, 1982. The interviewer had led Betty to believe that if everyone consented to participate in the interview session, that he would have a new chromosome test conducted at the Jet Propulsion Laboratory in California. It was claimed that the test would pin down the type of radiation involved and the extent of the resulting chromosome damage. Dr. Shenoy graciously agreed to support this worthwhile effort by making a public statement about the case. As an aside, the interviewer never made good on his promise to provide the chromosome test.

In response to questions about Betty's appearance soon after her arrival at the hospital, Dr. Shenoy said: "Miss Cash has been followed by us since 1979, ever since she had heart surgery. She was admitted to a nearby hospital on the second of January 1981. Her main problem at that time was a severe headache, a swelling of the face, a swelling of the eyelids, and a swelling of the skull on the top of the head.

He continued:

"Physical examination at that time showed a marked swelling of the eyelids. She also had marked swelling of the skull. One of the small glands in the back of the neck was swollen. We couldn't figure why she had marked swelling of the face, eyelids, and the skull.

Not knowing about the UFO encounter at the time, he thought she might be suffering from an allergic reaction of some kind and called in a dermatologist "to find out what exactly is the reaction to the face and to the hair."

Dr. Shenoy said the dermatologist saw her and commented on the marked reaction; but he was not sure it was an allergic reaction. He did find "a definite inflammation of the skull and the face."

Thinking the headache could be caused by inflammation of the sinus cavity, Dr. Shenoy ordered an X-ray of that area. The results were negative - "she did not have any inflammation of the sinus." He said: "Miss Cash did not have any headache in the past," so he called in a neurologist to evaluate the headache. Again, the results were negative. The neurologist said, "the headache is most probably due to severe tension headache."

Dr. Shenoy said: "Miss Cash responded well to the cortisone and the antibiotics." About the blood work, he said: "Most of the blood tests that we did in the hospital are negative for any kind of inflammation."

About her release from the hospital, Dr. Shenoy said:

"She stayed in the hospital approximately twelve days. The headache was getting better. The swelling of the face, eyelid, and skull was getting better. So, she went home..."

He also acknowledged why she returned to the hospital. He said: "A few days after going home she felt the headache is coming back again, the swelling of the skull is coming back again, and she noticed her hair is getting thinner and thinner. Four or five days afterward, she noticed she is losing hair in patches. So, she came back to the hospital... And when I saw her, I was surprised that the hair was gone on the head in patches, four or five big patches on the skull, measuring at least three or four inches in diameter. There was no damage in the skin." Dr. Shenoy again consulted with the dermatologist and found that he "couldn't pinpoint any definite cause for the loss of the hair."

It was at this point in time that Betty broke down and told Dr. Shenoy about the UFO encounter. The doctor was astounded and mildly upset that she hadn't told him earlier, as it could have altered his investigation of her problems. With this new information in hand, he resumed his search for the cause of the headache. He said: "I found that if she had an exposure to the intense light, she might have had damage to the cornea that is the front portion of the eye or to the retina, which is a screen in the eye. So, I sent her to the ophthalmologist. I told the ophthalmologist the story behind it and the reason for my concern. He took the history by himself, and he got the same story from Miss Cash. He examined her in detail, and he did not find any infrared damage to the eye or damage to the eye as a result of the radiation. He found a little bit of a problem with the lens. He could not document the damage to the retina either by the infrared or by the radiation.

Still puzzled by the hair loss, Dr. Shenoy said they clipped some hair and sent it to "the medical center" for examination. He said they wanted to find out if the medical center could document "whether the damage to the hair was from radiation, or this is due to the infrared, or whether she had some toxic reaction to heavy metals. Unfortunately, the hair they sent to the hospital was not fixed with the proper reagent, and they couldn't document the nature of the damage to her hair."

When questioned about whether or not Betty's current problems were related to her earlier heart problem, Dr. Shenoy said:

"Whatever symptoms she had, had nothing to do with the underlying heart problem. She had bypass surgery in Alabama sometime in 1977. She had a little bit of chest pain in 1979, at which time the cardiac cath showed one of the grafts was closing. At that time, we felt she was not a candidate for another bypass surgery, so we treated her with medicine, and she responded very well. She did not have any of the problems like high blood pressure, diabetes, or the tension headache. She responded very well to the pills we gave her for the heart. She has seen us every six months, and the problem is all gone. This is the first time I came across the swelling of the face and the intense headaches with Miss Cash. She did not have the symptoms in the past."

When asked how he reacted to her explanation of the UFO encounter, Dr. Shenoy said he made some telephone calls to physicians "who are functioning as radiologists, internists, and a gastroenterologist." He said they told him that Betty's problems were similar to those of other people who had encounters with UFOs.

He said: "Her story is reliable in that I trust what she tells me, and what she told me about the UFO. I don't think at any time I questioned her history. The neurologist, the dermatologist, and the ophthalmologist who saw her and got almost the same story as I felt that the story is reliable."

When asked if he has seen any other similar cases, he said: "I haven't seen any case which is similar to Miss Cash's problem, except the friend who traveled with her and the child who traveled with her in that same car, who had a similar reaction.

CHAPTER 19
Another Doctor Offers To Help

As soon as the investigation verified the existence of medical injuries, I began calling medical professionals in the state and federal government and in private practice. Nearly all were willing to listen to a brief overview of the case, but only a few were willing to engage in a discussion about what to do next and who else to contact. Some cited their job positions as the limiting factor in their getting more involved, while others were afraid, they might gel labeled as a kook, thereby hurting their business. In the business world, this situation is quite common, so I understand and respect their situation. One doctor, in particular, was an obvious debunker, and he was only interested in proving that the incident was a hoax. That was disgusting.

Then on March 12, 1981, I received an offer of help from a highly qualified radiation medicine professional. He was Dr. Peter Rank, Chief of the Department of Radiology at Methodist Hospital in Madison, Wisconsin. The timing was perfect. Under Dr. Rank's guidance, we were able to set up the protocol for obtaining medical records, as well as assessing the results of the treatments. His offer of assistance was especially gratifying to Betty, Vickie, and Colby. It came at a time when they were beset with trauma and didn't know which way to turn.

Dr. Rank provided a supply of medical information release forms which were quickly signed by both Betty and Vickie and submitted to their respective doctors and to Betty's hospital. By early April, we had received a substantial package of Betty's medical records. Unfortunately, however, none of the results of the blood work were included. Betty would battle with the hospital for more than a year before getting even one piece of blood work data. Her inquiries were always met with the excuse that the records were temporarily out of the file.

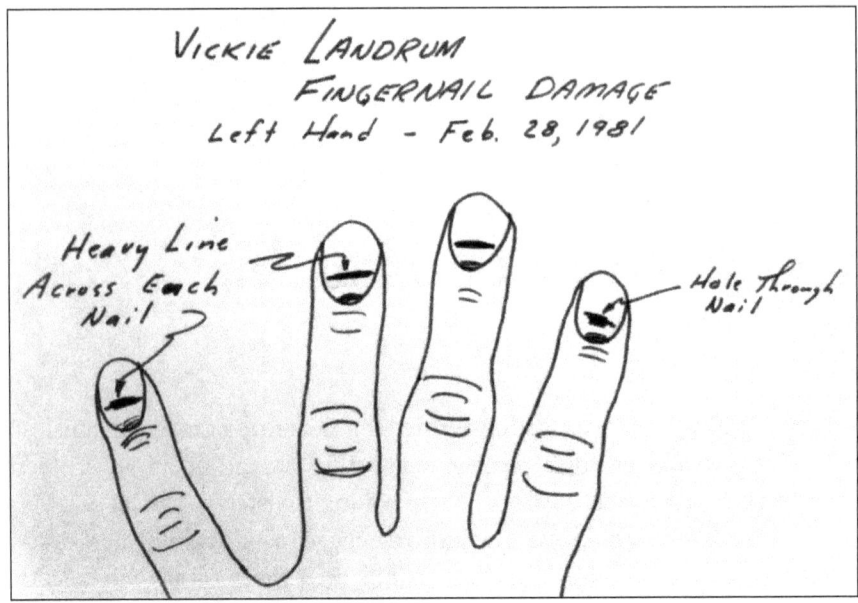

Sketch of Vickie's fingernail damage evident after the event, drawn February 28, 1981.

After reviewing the first package of Betty's records on April 12, Dr. Rank made the following observations:

1) An ophthalmologic and dermatology consult were included.

2) X-ray reports, EEG, lab reports, and pathology reports of a localized skin biopsy were also included.

3) There was no record of a complete physical examination available, and no progress notes.

4) The dermatologist's diagnosis was alopecia areata, a well-known benign dermis self-limiting affliction of the dermis resulting in a discreet localized loss of hair. Occasionally there is an underlying problem that is discerned as a cause. Most often, the cause of alopecia areata is unknown, and localized hair growth recurs. The dermatologist's opinion was that Betty had localized benign alopecia areata, and this was partially confirmed by the pathology report. I would caution you that this skin diagnosis is a wastebasket category and that Betty's loss of hair may not indeed be in that category.

5) The ophthalmologic examination shows nothing more than the

aging of the lens, causing difficulty with near vision. Fortunately, the ophthalmologist, Dr. Darsey, was prudent enough to mention that there was no evidence of damage to the interocular structures secondary to radiation.

6) There is no laboratory record available of white blood count studies, either a single or serial determination, and no record of platelet counts.

7) No discharge summary is included.

8) All x-ray studies are for the most part, non-contributory. The patient's heart is at the upper limit of normal, and there are scattered areas of fibrotic change in the lung fields consistent with the patient's age, not an unusual finding.

9) Please note that Betty has had a previous coronary artery bypass operation.

10) Electroencephalogram shows no abnormality and is therefore non-contributory to our current research.

11) They did make an effort to look for heavy metal intoxication as a cause of the patient's hair loss. According to the records supplied me this was an inconclusive study because insufficient hair was provided to the laboratory to make this chemical determination.

Dr: Rank commented on Betty and Vickie's fingernail damage. He said: "This shedding of the nails (onychomadesis) is a relatively rare condition that is brought out by the complete loss of growth of the keratin produced by the matrix at the base of the nail. It can be caused by a variety of situations, including severe febrile reactions, such as following scarlet fever. One, a few, or more nails may be lost. Such shedding of the nails also occurs secondary to emotional stress, is a psychosomatic expression of disease, although I think this unlikely. Most important of all, it is a result of severe local inflammatory changes and may occur in association with radiodermatitis secondary to ionizing radiation. Needless to say, the loss of hair is also consistent with radiodermatitis secondary to ionizing radiation. These two physical findings, specifically nail and hair loss, favor the presence of ionizing radiation."

After reviewing an additional package of medical records on April 29, 1981, Dr. Rank said: "I doubt the diagnosis of alopecia areata on many grounds. I will list these considerations as follows:

(A)Both women sustained some evidence of scalp damage, including hair loss.

B) Betty had a simultaneous loss of nails on the second through fifth digit of the right hand. All of Vickie's nails were affected.

C) The pathology report describes lymphocytes in the specimen, loosely arranged around specifically placed immature hair follicles. The hallmark of alopecia areata is not lymphocytes but mononuclear cells. Note that the biopsy was taken approximately four weeks after the original incident so that any acute white cells, specifically the polymorphonuclear cells, may at that time already be gone.

D) Her entire skin-scalp infliction began as cellulitis with considerable tenderness and erythema and responded to antibiotics. This is not a feature of alopecia areata.

E) The patient's entire history seems very reliable. She related the onset of her hair difficulties to the development of red, inflamed tissue, in turn directly related to UFO exposure. This is also not a history of alopecia areata.

F) The biopsy specimen describes the dissolution of intracellular bridges. This is not a feature of alopecia areata, but maybe a feature of ionizing radiation."

Dr. Rank provided the following information on the possibility of radiation exposure: "In regard to ionizing radiation, Betty and Vickie could have been exposed to alpha particles, beta particles, or gamma rays (including x-rays). I doubt alpha particles because while their linear energy transfer is quite high, they can be stopped in their forward travel by something as flimsy as a piece of paper. When they do attack the skin, their injury is limited to the most superficial keratinized layers of the skin. I also doubt beta particles for those same reasons, they could be stopped by a barrier as thin as the glass in an Erlenmeyer flask. Again, only the superficial layers of the epidermis would be affected by beta particles. We are therefore left with more penetrating gamma rays, which could penetrate not only to the subcutaneous layer but also throughout the body. By way of further elaboration, a neutron beam could also cause serious deep penetration. The hair *loss* in penetrating radiation occurs quite early. A dose of approximately 200 R to the skin is necessary for transient hair loss. Above 800 R hair loss is permanent. The growth of nails is disturbed in radiation exposure for the same reason that growth of hair is, which is that the germinal cells forming both hair and nails are affected by approximately equal doses."

"Whatever kind of radiation Betty and Vickie were exposed to; it was more penetrating than the most superficial types but still did not penetrate sufficiently to cause systemic signs and symptoms. It seems therefore safe to conclude, at this time, that Betty and Vickie sustained radiation damage which was confined to the skin and the immediate subcutaneous area."

"Whether this radiation damage was indeed due to ionizing radiation is at this point unclear. Other possibilities include infrared, ultraviolet, and microwave radiation. I know of no way that we will be able to establish anything further as to the type of radiation or dose. The conclusion I can make so far is that radiation damage occurred, type unknown, probably limited in severity, without evidence or systemic involvement."

During 1982 and 1983, Dr. Rank continued to review medical records as they were released and to answer my questions about the various noted conditions.

Betty, Vickie, and Colby experienced intense pain when they would take a hot bath and had to take cold baths. Dr. Rank said: "This symptom is well within the range of radiation response."

In discussing the problem of diarrhea, he said: "Obviously their bowel problems, including chronic diarrhea, may also be secondary to radiation effect. It may persist for days or weeks. Diarrhea may occur at doses in the area of 200-300 rads."

Their hair loss and regrowth is often seen in conjunction with radiation exposure. Dr. Rank said: "Hair does regrow following radiation damage with both a different color and a different texture. So far, the outcome of radiation epilation cannot be predicted. Epilation may occur anywhere from a few days up to a year. This is based not only on the literature but my experience in radiation therapy, as well as a personal experience in my family."

About the radiation effects on the skin, he said: "The dermatologic manifestations of radiation depend upon the wavelength and the total dose and may be either acute, chronic, or a mixture of both. Some immediate post-radiation dermatitis begin within hours of exposure. Others do not develop for 20 years. An acute radiation dermatitis may result in the non-healing ulcerative lesion, and it may not necessarily indicate an exposure of fatal proportions."

Dr. Rank spent a lot of time telling me about how different individuals react to the same radiation exposure. He said that it seems that their reaction is affected by their diet, size, drugs taken, and even the amount of caffeine consumed. The problem we have is that radiation creates a lot of broad nonspecific physiologic effects that are determined not only by the exciting agent, but more importantly, by the limited ways that the body can respond to stimuli.

CHAPTER 20
Hot Air Balloon Not The Culprit

Since the object was huge and flames were involved, we decided to investigate the possibility that the UFO was actually a hot air balloon. We recognized at the start that most hot air balloons are 'pear-shaped,' not diamond-shaped; however, there are specialty balloons resembling everything from the space shuttle to big boxes. The investigation was fairly extensive, but the results were negative in every case.

Initially, we focused on the verbal descriptions given by Betty, Vickie, and Colby. Could they have seen a hot air balloon and, due to the darkness, thought it was something else?

They pointed to their sketches and again told how the flames shot down toward the ground and not up into the object. They told how bright the object appeared— so bright and hot that it hurt their eyes and burned their skin. They are sure they were not seeing a balloon illuminated from the inside by the flames from the gas burner. They pointed out how the noise from the flame was so loud that it was not like a hot air balloon. They were adamant that the object they encountered was not a hot air balloon.

The route taken by the object also ruled against it being a hot air balloon. It approached the area from the east, hovered over the road for a period of time and then flew away in a southerly direction. This would be a difficult, if not impossible maneuver for a hot air balloon.

The weather at the time of the incident would probably not have stopped a daylight flight of a balloon, except for the fairly swift wind that was blowing. Late in the afternoon, however, at the time balloons are usually flown in the Houston area, the experts said the weather was unsatisfactory for balloon flight.

A spokesman for the Houston Intercontinental Airport said there were no balloons in that area on December 29 and none are ever allowed at night.

According to the hot air balloon experts and several independent helicopter pilots, the fact that the object was intercepted by helicopters at night while in view of the victims and then flew away from the area with helicopters flying around it rules out the flight of a hot air balloon. All say it would not happen like that.

Bill Murtaugh, head of the Balloon Pilots Association, said no balloons were certified for night flights, and none would fly in that segment of the greater Houston area because the terrain is much too rough and too close to the big airport.

John Wyler of Balloon Adventures said: "No balloon in Houston is certified to fly at night."

Similar answers were received for representatives of the other companies in the Houston area, including Air Adventures, Lone Star Balloons, Pretty Balloons Unlimited, Raven Hot Air Sport Balloons, and Texas Balloon Stables.

CHAPTER 21
Lack Of Radar Hampered The Investigation

The UFO and the helicopters may have been spotted and tracked by radar. The problem comes in proving that assertion. It was approximately two months after the incident before the investigators made the first requests for radar data. An Air Traffic Control Center spokesman told our investigators that all radar tapes are pulled and destroyed after 30 days. Therefore, the tapes were no longer available for analysis. The spokesman also said he did not recollect anyone mentioning the high level of activity on December 29, 1980. However, he was off duty during that part of the holiday period.

It is likely that the tapes would have shown nothing because the radar is blind below an altitude of 2,200 feet in the area where the Cash-Landrum encounter occurred. The UFO and the helicopters operated well below that altitude throughout the duration of the incident.

Low altitude helicopter operations are common around Houston. For example, the Army's 136th Transportation Unit operation CH-47 helicopters out of Ellington Field on the southeast side of Houston flew across to the northwest side of Houston on a regular basis for operations at the Addicks Reservoir, and they did it without going through Houston Air Traffic Control or being spotted by radar.

Requests to official agencies for information were futile. No agency or organization would accept responsibility for the helicopters or the UFO; therefore, they refused to dig deeper for data that could have been helpful.

Official radar records are not available for general scrutiny in any case. Therefore, the investigation was inconclusive in proving whether or not radar records existed or if they showed evidence of the Cash-Landrum encounter.

On the positive side, our investigation into how the activities of airborne drug smugglers are monitored along the Gulf of Mexico coastline and the United States/Mexican border provided some clues that were useful in analyzing the UFO situation.

According to *The Houston Post*, dated April 28, 1985, in the early 1980-time frame, the United States Air Defense Network was focused on the areas to the north of the United States. The only worry to the south was about high-altitude devices. Ground radar of the North American Air Defense System (NORAD) could not detect low-flying vehicles of any kind on the southern approaches, not drug planes, helicopters, or unidentified flying objects.

Texas had no fixed-radar coverage in the 500 miles between Laredo and El Paso, or at key points along the western Gulf coast. There were extensive gaps in low-altitude radar coverage along the border, ranging from 2,500 feet to 5,000 feet, and for several hundred feet, this gap extended upward to 14,500 feet. Pointing to the holes in the radar coverage, Tom Bailey, chief of the U.S. Customs Service's air support branch in San Antonio, said: "We frankly don't know how many illegal flights there are across the border. I've heard estimates of 10 flights daily, and I've heard 150." If they were unable to see airplanes, then they were also unable to see UFOs.

Texas Governor Mark White said: "I can't help wondering why our military is so worried about stopping a future invasion of Russian bombers along our northern border when it is doing nothing to stop DC-3s loaded with drugs from entering this country from Mexico."

Jim Adams, Director of the Texas Department of Public Safety told Congress: "You'd have to try real hard to get caught." He said they probably catch less than one-tenth of the traffic funneling through the gaps.

Kay Cormier, spokeswoman for NORAD Headquarters in Colorado Springs, Colorado said: "We realize our air defense system is archaic, outmoded, and outdated. It was fine 20 years ago when we feared a high-altitude bomber threat more than a cruise missile threat."

Today, improvements have been made. Holes in the southern radar fence have been plugged by Over-the-Horizon Backscatter radar, a system that bounces a radar signal off the ionosphere back toward Earth. A PAVE PAWS phased-array radar system went into operation near San Angelo in 1987. None of these capabilities were available in 1980.

The first line of defense along the Texas Gulf Coast is the 147th Fighter-Interceptor Group of the Texas Air National Guard, stationed at Ellington Field. They react to instructions from Tyndall Air Force Base in Florida. When Tyndall says there is something over the Gulf, this group investigates and intercepts. Low-level flyers aren't noticed.

One possibility for low altitude coverage along the Gulf coast was the Air Force Airborne Warning and Control (AWACs) aircraft. These planes look down from above and can spot low flyers. Their success rate for catching drug traffickers was pretty poor. Texas Governor Mark White said that in 1984 the AWACs used for drug surveillance along the Gulf coast failed to lead to a single arrest after 500 hours of flying time.

Many more examples of the holes in the southern radar net could be cited, but that is not necessary for this investigation. The lack of low-altitude radar coverage, coupled with the large number of UFO sightings on December 29, suggests a scenario that has the UFO enter the United States by flying over the Gulf of Mexico at low altitude, crossing the coastline between Morgan City and Lake Charles, Louisiana, or between Lake Charles and Beaumont, Texas. Once inland, it turned westward and moved slowly to the point where it hovered over the road in front of Betty, Vickie, and Colby. From that point, it curved to the southwest and finally to the south to follow the river and fly over sparsely settled areas to eventually cross the coastline again, this time near Baytown, Texas.

Using the same radar coverage information, other flight path scenarios could be defined. However, the large number of visual sightings of the UFO, plus the helicopter activity, coupled with the radar coverage information, tends to support the coastline-crossing scenario

The lack of radar data has hampered the investigation of this case. In particular, good radar coverage could have shown beyond a shadow of a doubt where the UFO came from, where it went, and the extent of the helicopter activity.

Knowing that the Texas Gulf coast was almost totally lacking in radar coverage does explain how the UFO could operate almost without official detection regardless of the origin of the UFO - from the United States, from some foreign power, or from outer space. Even if there was official detection of the UFO, the lack of radar coverage scenario provides a good excuse for the government to plead ignorance in this case.

> **WIN A TRIP TO WASHINGTON D. C.!!!**
> TRAVEL ARRANGEMENTS COURTESY OF DELTA AIR LINES
> AND DUPONT PLAZA HOTEL
>
> **2ND ANNUAL**
>
> **McDonald's®/Houston Chronicle Children's Creative Writing Contest**
>
> Write a story on any subject you choose. Five (5) of the ten (10) words below must appear in your story. Underline the five (5) words where they appear. The words must be used as they are shown, the tense cannot be changed. You may select only five (5) of the words and your story must be a minimum 25 words.
>
> PILLOW RUNNING RAISINS NEWSPAPER SMILE BURN
> ERASER FENCE ROCK BUTTON
>
> Name: Colby Landrum Phone: 258-8709
> Address: 44 Brown Road Age: 7
>
> The Nightmare that won't end
>
> The nightmare began on a monday night on Dec 29 1980. my grandma and my aunt Betty and me went to new Caney Texas and on our way back it began to happen. I saw a bright object in the sky as we went down the dark road it grew closer and bigger and brighter. It was diamend shaped and fire was coming out of it. As it hung above the treetops we had to stop. I tried running away but had to get back in the car

Copy of Colby's entry into the Children's Creative Writing Contest, sponsored by McDonald's and the Houston Chronicle. Colby chose to write about the incident.

> Name _____ Phone _____
> Address _____ Age ____
>
> I turned red from the heat, did it ever burn. We had blisters aunt Betty was the worst. We all was so sick we lost all our hair to. The newspaper tried to help us find out what it was But we got no help. Its hard to smile sometimes when I remember. Sometimes I wake up crying in my pillow. there was helecopters that night with big Blades, maybe someday I can find out what was and the nightmare will end. This is a true story. I was there
>
> The END

Continuation of Colby's story.

CHAPTER 22
Troubled Times

The year following the incident was a challenging and difficult time for Betty, Vickie, and Colby. At times they felt like they would not survive. To make matters worse, their continuing poor state of health made it difficult for them to pursue their quest for information about the UFO and the military helicopters that chased it.

In between their bouts with nausea, vomiting, open sores, and pain, they contacted governmental agencies, politicians, and medical help groups. In every case, they openly offered their medical records and their personal testimony. Some of the agencies treated them with respect and led them to believe they would be helped in some way, a promise that never materialized. People in other agencies were just plain rude, at times leaving them in tears after their calls. It was very much like the way victims of Agent Orange and the Gulf War Syndrome were treated.

When all else failed, we carried their plight to the media. It was our hope that the media contact would help to force the government agencies to act more responsibly by actually looking into their case and that it also might convince some of the special forces personnel in the helicopters to come forward with information that would be helpful. As will be documented in later chapters, we were somewhat successful in opening a dialogue with a few of the governmental agencies, but not with the helicopter flight crews.

I was in contact with Betty and Vickie weekly, if not more often, throughout the year. Since the volume of my recorded investigative notes for 1981 is sufficient to fill a book, I have extracted only some of the conditions and activities following the initial investigation that show how the situation completely disrupted their lives.

During the year, Betty and Vickie were bombarded by inquiries from newspapers, tabloids, radio, and television. MUFON, APRO, and CUFOS, which were the leading U.S. UFO investigative organizations at that time, all sought information for their respective journals. Late in the year, Betty and Vickie did finally consent to a television interview with "That's Incredible," and their plight became known to thousands of television viewers.

Betty's home in Dayton, Texas.

Since Betty was too ill to function without assistance, her mother, Pauline Collins, came to Houston to care for her after Betty left the hospital the second time. Mrs. Collins said it was easy to see that Betty was not going to have a speedy recovery. Therefore, on March 23rd, she took Betty back to her home in Birmingham, Alabama, where it would be easier to care for her.

When she left Houston, Betty was feeling poorly. Her eyes were badly irritated, and she was suffering from headaches, lethargy, and weakness. Her hair had grown very little, so she still had large bald spots on her scalp. And blisters like those originally covering her face were now appearing on her arms, legs, and back.

The doctors in Birmingham started what would be many years of care and treatment. During the next period of hospitalization, they determined that Betty's skin tissue was breaking down. After verifying that the condition was not caused by fungus or allergy, the experts concluded it was due to radiation exposure.

During that same month, Vickie and Colby were equally troubled. Vickie was still sick, and her burned areas were covered with sores. Her hair was also slow to regrow, and her eyes were badly irritated. The headaches continued unabated. Her doctor was unable to give her any relief.

Colby's eyes were still very irritated. He had stomach cramps as the diarrhea continued. And he still had skin eruptions. Because of his condition, he had no appetite, and he suffered from weight loss. Worst of all, however, he suffered from continual, horrible nightmares.

During the month of April, Vickie made weekly visits to a local clinic because she was feeling sick and dizzy like she was about to faint. She seemed to be getting weaker and weaker. Historically, she always had low blood pressure. The first week her blood pressure was 160/87. The second week it was 142/84. The third week it was 120/80, and it leveled off at 110/80, still very high as compared with her past records.

Colby's condition continued much as it was during March, except a large water blister grew on the side of his face, and after a few days, it broke, flowing a clear liquid. Vickie said it looked really bad. And his nightmares continued.

Betty's condition did not improve. To make matters worse, she began having nightmares.

It was during this period that Betty and Vickie each wrote to their Congressmen and Senators. Congressman Charles Wilson didn't reply until July, and then all he did was recommend that they contact Ufologist Frank Stranges. Soon after, Senators Bentsen and Tower sent letters telling them to go to Bergstrom Air Force Base at Austin, Texas for help.

Throughout this period, Betty and Vickie were continually reminded about their exposure to radiation by the large black line across their fingernails and the holes through several of their nails in the middle of the black line. By the month of May, their nails had grown sufficiently that they could begin to clip off this unsightly reminder of their encounter.

Vickie Landrum's eyes were damaged during the incident and she continued to have sight diffculties.

Nothing changed much during May and June. Betty continued to feel bad and suffer with the blisters. Vickie's eyes were so bad by this time that her glasses were useless. Colby still has the stomach problems and the nightmares. At the end of May, Betty joined Vickie at a hospital in Houston to have the sores examined, mapped, and photographed. Nothing ever came of the hospital work-up.

During the third week of May, Vickie received a call from a person claiming to be from the FBI. He told her they would be sending two agents out to talk to her. Since it seemed unusual for the FBI to send agents to investigate a UFO incident, I called the Houston FBI office where I spoke to Agent McGee. Ms. McGee had no record of agents being sent to see Vickie. She said it was "possible" that they would send an agent on this type of case, but it "was not probable." Ms. McGee told me to have Vickie call the FBI at 224-1511, a number available 24-hours a day if anyone claiming to represent them came to her home. She also said to get their full identification, including names, numbers, where their I.D. cards were issued, telephone number, auto type, and license number. She said, "to take no chances." The agents never arrived. Whether it was because I called the FBI office and questioned the visit or because the call was bogus, we will never know.

Betty's house in Dayton had been closed up since she went to her mother's house in March. She was shocked to find out it had been robbed. Strangely, the robbers opened all of her closets, drawers, and storage boxes and "ravaged" everything. Nothing of value was stolen. The only things missing were personal items, papers, and records.

Colby began having dental problems by mid-1981. This is unusual because he never had a problem before. In July, he developed three cavities and many more in the latter half of the year.

Vickie underwent regressive hypnosis on July 11. She was terrified as she relived the event. In the weeks following the hypnosis session, Vickie was upset and ill. Instead of helping, the session seemed to make her worse. She said: "I'll never relive that terror again."

On July 19th, Vickie heard about three other women that were also burned by a UFO on December 29, 1980. She called me in an excited state, believing that she had found additional witnesses to the event. We tried to locate the women by contacting the area hospitals, but to no avail. The hospitals said they couldn't reveal information about patients. When that didn't work, we contacted the area police and sheriff's offices. Office personnel all stated that no women had made a police report. This was not surprising because they also had no report of the Cash-Landrum incident.

No one had called the police that night either.

Betty, Vickie, and Colby, obedient to their government leaders, did follow the instructions received from Senators Tower and Bentsen. They drove to Bergstrom Air Force Base on August 17th and told their story to the Air Force lawyers. Unfortunately, after giving their testimony for the record, they received no assistance from the Air Force. It was a disappointing day. Their feelings of trust and patriotism had been shattered because their government had let them down.

August 20th was a bad day for Vickie. She said she had no vision in the left eye. It was like she had a patch over it. It was badly swollen. This was a shock to Vickie because her left eye had been the best up to this point in time. By August 23, she had regained some very diminished vision in the left eye. Her eye doctor referred her to another specialist on August 25th. He said she now sees out of a very narrow cone. He surmised she might have optic nerve damage. At that point, her eye doctor felt that he could do no more for her and gave her a referral to the Ocular Diagnostic Clinic at the University of Houston. When she went to the University of Houston on September 8th, they told her that the back of her eye was fine, but her eye had damage around its edges.

By the end of August, Colby was wearing glasses during the day. He was still suffering with stomach aches and had periodic bouts with loose bowels. He was still suffering from a weight loss condition. His size six slim jeans that fit loosely in May now fell off his hips. He couldn't keep them up. The good news was that the small bald spot where the hair loss had occurred was now filling in nicely. Strangely, however, he had grown some unusual clumps of hair on his back and arms during the summer. The clumps ranged in size from one-half to three-quarters of an inch in diameter. On August 29th, Vickie caught him trying to shave the clumps of hair to get rid of them.

Vickie said Colby was having trouble eating. When he did eat, he got stomach aches and had bowel problems. Even when given a wide choice of foods, he still would not eat. He started the day by eating only a part of a bowl of cereal. Prior to the UFO event, he liked coke and hamburgers. Afterward he wouldn't touch either one. He did drink a lot of water. He seemed to have an unquenchable thirst. Betty and Vickie also had the problem of thirstiness.

Vickie was dealing with another kind of hair problem. After her hair loss, the hair that grew back was of a different color and texture. It was very coarse. This was bothersome while it was short, but when it grew longer, it was easier for her to manage than before the hair loss. In the end, she liked the new hair better.

Betty made several trips to the hospital during 1981 and spent some time in cardiac care in September. When her mother called to tell me that Betty was in the hospital on September 30th, she was crying because she was so worried about Betty. I called Dr. Whittaker that same night, and he told me Betty did not have a heart attack. She had a cold, some angina, and was spitting up. He said he would release her from the cardiac unit the next day.

One of Betty's most dangerous sessions started on October 5th. She had experienced a long period of weight loss and was down to 103 pounds. Then she developed chest congestion, cramps, diarrhea, and chest pains. The large blisters were again breaking out on her body. Dr. Whittaker was concerned and ordered an extensive battery of tests, including a lot of blood work. When I spoke with Betty, she said: "It might feel better to be dead." I reminded her that she needed to continue the fight for justice and that I would help her in every way possible.

Betty contacted me after she was finally released from the hospital again because she was concerned about some of the things she was told while in the hospital. Dr. Whittaker called in some other specialists because Betty was not responding to antibiotics. They were interested in her description of the original event and the possibility that she had been subjected to radiation exposure. They said something was blocking the antibiotics from curing the infection, and it had them all stumped.

Blood work done on Betty, Vickie, and Colby in November showed they were all still off nominal in red and white blood cell count.

On the 28th of November, Betty went back to the hospital believing she was having a heart attack and was admitted to intensive care. They noted her condition to be "serious." She had swollen legs, a lot of skin eruptions, and low blood pressure. Tests showed she was not having a heart attack. Instead, she was suffering from an inflammation of the heart sac, referred to as pericarditis. It was at this point that Betty's doctor was quoted as saying her condition was "secondary to radiation exposure."

Betty's doctors were excellent caregivers. They successfully helped her to survive her many life-threatening medical events. And unlike the problems Betty had in getting copies of her original hospital test results, these doctors openly shared the records of her treatment with her. These records are intact and available if Betty needs them in the future.

The problems experienced by Betty, Vickie, and Colby were nicely stated by Colby in his own words when he submitted an entry into the Children's Creative Writing Contest, sponsored by McDonalds and the Houston Chronicle. This is what he said:

THE NIGHTMARE THAT WON'T END

The nightmare began on a Monday night on Dec. 29, 1980. My grandma and my aunt Betty and me went to New Caney, Texas, and on our way back, it began to happen. There was a bright object in the sky as we went down the dark road. It grew closer and bigger and brighter. It was diamond-shaped, and fire was coming out of it as it hung above the treetops. We had to stop. I tried running away but had to get back in the car. I turned red from the heat, did it ever burn. We had blisters. Aunt Betty was the worst. We all was so sick. We lost all our hair too. The newspaper tried to help us find out what it was, but we got none. It's hard to smile. Sometimes when I remember, I sometimes wake up crying in my pillow. There was helicopters that night with big blades. Maybe someday I can find out what it was, and the nightmare will end. This is a true story. I was there.

The End

None of the investigators or media people knew Colby was entering this contest. While he didn't win the trip to Washington, D. C., this seven-year-old did win our hearts, so we all renewed our efforts to help him end his nightmare.

CHAPTER 23
Hypnosis Was Used

Colby's nightmares were very real, and Vickie often expressed her concerns about his nightmares, wondering what she could do to help him. Her concerns increased when she heard Betty was having nightmares also.

Betty's dreams were very traumatic. For example, her sister woke her from a horrible nightmare during the night of April 22, 1981. She said Betty was crying very hard and screaming, "Go away! Leave me alone!" In describing the dream, Betty said she was by the UFO, and some weird-looking beings were after her. She described the beings as being human-sized but more ape-like. As they came for her, she felt paralyzed and could not get away. Her screaming was a cry for help.

The intense nature and increasing frequency of their dreams suggested that more information about the incident might exist. Perhaps there had been beings on the ground near the UFO that night. Perhaps the paralysis was a real event and was only being revealed through the dreams.

After consultation with experts in the field, it was decided to try hypnosis as a tool for recovering hidden information from the witnesses. Betty was willing, but her doctors didn't think she was strong enough to relive the event again. Colby was too young and vulnerable. That left Vickie to carry the burden of going through the regressive hypnosis sessions and experiencing the trauma again. Because of her worries about Colby, Vickie readily volunteered.

The next step was to select the appropriate hypnotherapist from the field of qualified caregivers. Since our primary interest was the well-being of the victim, we wanted to use a professional with a proven track record in this type of case. After an extensive search, it was concluded that the most caring and loving professional in the field at that time was Dr. R. Leo Sprinkle, Counseling Psychologist at the University of Wyoming at Laramie. Dr. Sprinkle accepted the assignment, and the date of the first session was set for July 11, 1981.

The session was conducted in the living room of Vickie's home in Dayton, Texas. Vickie was so concerned about having to relive the event that she said, "I think I will die if I have to." Dr. Sprinkle eased her concerns by offering her the option of "seeing it on a screen" as an alternative to reliving it. She agreed, and the session proceeded.

There is no reason to repeat the whole session in this chapter, but the parts of the session that relate to new information such as talking about the experience, odors, lights, and the possibility of this being an abduction event will be stated verbatim.

THE INITIAL EXPERIENCE: Dr. Sprinkle proceeded to place Vickie under hypnosis, while assuring her she would be able to tolerate any fear and pain. Then he said: "You can let yourself go back and talk about your experience. Whenever you're ready just go right ahead and talk about your experience."

Without further interruption from Dr. Sprinkle, Vickie proceeded to talk through the various phases of the experience, starting with the initial encounter: "O Jesus. He's coming. See it. See it. It's like a diamond...."

"Colby, Colby...(unclear).... Jesus won't hurt you, Colby. He's coming back, just, just watch. Momma's going to get back in the car. Momma loves you, darlin.'"

"Betty! Betty! (Shouting) Come! Come! Come right to this car! Don't walk toward it! Oh, Betty, please! Oh, we're gonna burn up, Betty, oh we're gonna burn up. I know we are. Oh, Betty, please...."

"Colby, just don't be afraid. Mommy loves you better than anything. But that's Jesus. He's coming out of the sky. When you see this big, big man, you're gonna know he's Jesus. And he's comin' after us, and he will not hurt us. There is no way that God ever hurt anybody. I know that this is it. God, I have done my best! So, please take me as I am. Jesus, you know that I have really tried. I lived the best I can. Jesus, please, take care of my baby, for he's little. We'll go by and get poppa. Don't worry; we'll pick him up. We're going to carry him with us."

"It's burning! Look, the whole thing is burning. There's fire coming out the bottom. It let up. I think it's on fire.... no....thank God....It's.... it's lifting. But I think my lighter's leaking."

"Just look at all them helicopters. And they're not the little bitty kind. I know, Colby, there's double rotaries there. I know it darlin'.... just don't worry about that. We're safe, and we're sound, but I'm burning, and it's so hot. Don't seem like the air conditioner's doing no good."

"Betty, don't stop here. We can see 'em from the car. I don't know how many I counted. I think I counted 20- 21- maybe 23.... Look at all them helicopters."

In her description of the events that occurred later that night when she confronted her own and Colby's injuries, you could hear the fear and pain in Vickie's voice as she talked to Colby:

"We can't tell nobody! But I'm burning up! Lord help me! Oh, that water feels good. Mommy's just going to put some oil on you. That oil will help you a lot. Then you can go to sleep. Please don't tell nobody. 'Cause they'll think we're crazy. My stomach hurts. My head hurts. I'm sick. Colby, Grandma can't help it!"

Then Vickie talked about trying to help Betty on the following day: "Betty, Betty, Betty, honey, what's wrong? I'll get you some water, and then I'll carry you to my house. I want to take care of you. Lord, I don't know what I'm gonna do! I really don't know what I'm gonna do! Betty, is there anybody we can call? Eyes are hurting. Oh, I hurt (crying), I hurt. Can't stand it! My eyes! I gotta get you to the hospital .... Doctor, I need to make an appointment for Betty Cash. It's an emergency. Dr. Wilson would you possibly see her? Why not? I know that, but I need help. Betty, I can't get nobody. Dr. McDonald, I need help. Can't you just see her? Betty, I got to do something. I called the drugstore. I got a number. Oh, Betty, lean, get you in the hospital! Thank God! Wilma, I think Betty is dying. Would you come and help me? Louise, would you please drive Betty to the hospital?"

As Vickie continued to relive the events, she told her husband what had happened to them for the first time:

"Darling, I've gotta tell you something, and I hope you believe me. We run into something bad, and Betty's dying, and Colby's got the virus, he's wetting the bed, he's having nightmares. So, whether you believe me or not,

I don't think I'm crazy. We run into something. Were no little green men come out in a UFO, 'cause it was fire, lots of fire, and it was hot, and I just put Betty in the hospital, and I don't think she'll make it. I'm glad you believe me 'cause I never lied to you before. I know you love me. But I can't cook, 'cause I'm burned too bad. I know you will... I can't find out what's wrong with Betty."

Vickie finished describing the event by talking about the pain in her eyes, saying: "My eyes! My eyes! My eyes!" So, Dr. Sprinkle tried to help her ease the pain, but she said: "It don't ease; they stay hurting all the time."

Next, Dr. Sprinkle told Vickie to revisit certain specific events from the encounter and began to ask specific questions. In response, Vickie described where each of them had been sitting in the car as they approached the object, where they stood after the car stopped and they got out, how many helicopters there were and how some were different, and how she got back in the car to comfort Colby. Then, she described where the object was located, how it sounded, and a special odor that she noticed, as follows:

(Q) And where is Betty?
(A) Outside
(Q) She's outside? Can you see her now?
(A) Uh-huh.
(Q) She's standing by the car.
(A) Uh-huh
(Q) By the door?
(A) Uh-huh.
(Q) And what's she doing?
(A) She's showing Colby.
(Q) She's what?
(A) Showing Colby.
(Q) She's showing Colby.
(A) Uh-huh.
(Q) She's pointing somewhere.
(A) Uh-huh.
(Q) Where is she pointing?
(A) Up this way.
(Q) Up this way? Pointing at the object? Helicopters?
(A) Uh-huh. And the thing.
(Q) And the thing?

(A) 'Cause it didn't go away.
(Q) It's still there?
(A) Uh-huh.
Q) At about what angle is it?
(A) 'Bout like that (pointing).
(Q) In front of the car?
(A) Uh-huh.
(Q) Any sound?
(A) Uh-huh.
(Q) Yeah, what kind of sound?
(A) Sound like a tornado.
(Q) Sound like a tornado? Really loud?
(A) Uh-huh.
(Q) Any smells?
(A) No, nothing but my lighter.
(Q) Your lighter?
(A) Smelled my lighter.
(Q) You smelled your lighter. As if the fluid was coming out?
(A) Uh-huh.
(Q) Did you, did you see it, did you notice it, your lighter?
(A) It wasn't (there) when I got home.

Following the hypnosis session, we discussed the lighter smell at length and found that Vickie clearly remembered the smell of lighter fluid or a kerosene-like odor. The investigation revealed that neither she nor Betty had used a lighter with lighter fluid in a long time. They were using a butane gas lighter, and the smell was not the same. Our conclusion from that fact led us to believe she had smelled the exhaust from the helicopters.

The next area of inquiry was the heat. Dr. Sprinkle wanted to see if the heat level was consistent with their earlier testimony, so he probed for more information:

(Q) How about, lotta heat did you say?
(A) Uh-huh.
(Q) How much?
(A) Hot! Hot!
(Q) How hot is it? Can you estimate how hot it is?

(A) It is hot enough that some sweat was coming out.
(Q) Sweat was coming out?
(A) Uh-huh.
(Q) Ok, what else did you notice? Just look around you and see if you notice anything else.
(A) Just when it would let up is when it would come down. When the fire'd come out.
(Q) It would let up.
(A) It would go back up.
(Q) The fire would go back up.
(A) No!
(Q) No. Oh, the object would go...
(A) Well, when that fire... I can't explain... fire would seem to go out of the plane. It would go up, and then it would come down, and it would come down, and it was kind of a rocket fire.
(Q) Like rocket fire coming down.
(A) Uh-huh. Like when they shot off that missile there, you know how the fire comes out.
(Q) Uh-Huh, fire comes out of the missile. That's the way it was.
(A) There was fire in the bottom of it. It lit the whole thing up—hot, hot fire.

Finding the heat level to be consistent with earlier testimony, it was time to seek more information about the size of the object: Dr. Sprinkle continued:
(Q) And how big does it seem to be?
(A) Mmmm, it was bigger than a water tower.
(Q) Bigger than a water tower? Like 100 feet or 200 feet or 50 feet?
(A) About as long as a rocket, I think. Maybe not that long.
(Q) About as long as a rocket, but maybe 'bout as tall as a water tower?
(A) At least.
(Q) At least as tall?
A) Uh-huh.
(Q) And the shape?
(A) It was shaped like a diamond with the point cut off the top of it.
(Q) A diamond with the point cut off of it. And which way does the shape face? The point, the diamond, the point cut off—is that facing up, or

sideways?

(A) It was, uh... fire coming out of the bottom.

(Q) Fire coming out of the bottom. And the broken point or the blunt point up on top?

(A) Uh-huh.

(Q) Could you see how wide or how tall?

(A) It was long.

(Q) It was long?

(A) Kind of wide.

(Q) Kind of wide? About like a diamond, maybe a thin diamond?

(A) Thin diamond.

(Q) Uh-huh. Could you see anything else, any other features, or...

(A) It had some blue on it.

(Q) It had some blue on it.

(A) Uh-huh.

(Q) What kind of blue?

(A) It Looked like little lights.

(Q) Looked like what?

(A) It Looked like little lights.

(Q) Like little lights. Like little blue lights.

(A) I think so.

Later Vickie explained how the row of little blue lights was located around the mid-point or widest part of the diamond shape.

Dr. Sprinkle continued to probe and found that Vickie could describe a sound that was present during the encounter, as follows:

(Q) OK. Good. Anything else you noticed?

(A) It beeped.

(Q) It beeped? What kind of what kind of beeps? Can you make a sound like the beep?

(A) Uh uh.

(Q) Hard to do, huh? Can you describe what kind of sound? You heard anything like it, similar to it? What would be the closest sound?

(A) There ain't one.

(Q) Say again.

(A) There ain't one.

(Q) Tell me again.

(A) There ain't one.

(Q) Oh, there ain't one. There's no sound like it?

(A) Yeah.

(Q) OK. Was it a high-pitched beep, or was it low-pitched?

(A) Kinda high.

(Q) Kinda high. And was it a short beep or a long beep?

(A) Kinda in between.

(Q) Kinda in between. Like maybe one second or two seconds? Is it like this: beep.... beep....beep....or is it faster?

(A) Longer.

(Q) Longer? Was it beeep ...... beeep ...... beep ......? more like that?

(A) No.

(Q) No, not that high. Beep .... More like that?

(A) You can't do it the same.

(Q) I can't do it, not even close, eh? But if you had an instrument that could make that sound, would it be regular like boom, boom, boom, or would it be irregular boom.... boomboomboom....boom?

(A) It was irregular.

(Q) OK, so it was really hard to know what kind of sound, what kind of a thing would make that sound. Did it sound musical-like, or did it sound noise-like?

(A) Noise.

(Q) Noise-like. Did it sound metallic, like metal?

(A) There's such a roar I don't know.

Dr. Sprinkle gave Vickie permission to relax and be free of pain and free of fear while he was determining whether or not to ask more questions. Suddenly, Vickie went back to talking about the odor:

(A) I keep smelling lighter fluid.

(Q) You keep smelling what?

(A) I keep smelling lighter fluid.

(Q) You keep smelling lighter fluid? OK, see if you can notice where the smell comes from. Is it from your own lighter, or is it from some other source? Can you tell? Can you tell if it's from the car or outside the car?

(A) Don't remember.

(Q) Don't remember? Just look around and see it seems stronger in one direction, or it seems stronger in another direction. It's like in your mind's eye, you're just sniffing to your left, to your right, in front of you, in back of you—just circle around and see where the strongest scent seems to come from. Does it come from above or below? The back seat, the front seat, or outside the car?

(A) I don't know.

(Q) Hard to know where it comes from? Not only use your nose but use your mind. Use your mind's eye, almost like you are beaming in on it. See if you can track it down. See if you can tell where it comes from. Let your imagination, let your mind be aware.

(A) It comes from up this way (pointing upward).

(Q) Up from there. Good.

(A) Cause we saw it.

(Q) That's when you saw it, that's when you smelled it?

(A) No.

(Q) The smell later.

(A) We saw it down the road going down, down meeting it.

(Q) You mean it was after the object, or before?

(A) Yes.

(Q) Before the object appears, you notice the smell before the object appears.

(A) No.

(Q) No? OK. Tell me again.

(A) I said we saw the object, and we go on and on...

(Q) Uh-huh. Go on and on... When you're driving, were you going on and on?

(A) Uh-huh.

(Q) Yeah. Is this when you first saw...

(A) And then it stopped.

(Q) She stopped. Betty stops the car.

(A) I told her to.

(Q) Uh-huh. Ok. Betty stops the car. Now what happens? Just go on through.

(A) We would 'a burned up.

(Q) If you'd go too close, you'd a burned up?

(A) Yeah.

(Q) That's when you noticed the smell?

(A) No.
(Q) No. That came later.
(A) And the fire was coming down.

Dr. Sprinkle decided it was time to seek more information about the helicopters, such as: when they first noticed the helicopters, how many, if other airplanes were present, and how they flew around the object.

(Q) One question comes to mind. When you and Betty and Colby are there, and the car is stopped, and now you're outside the car, and you're looking up at the object. You're looking at the object, and then after a while you notice helicopters coming in. When do you first notice the helicopters—about how long between when you first see the object, and you first notice the helicopters? Can you estimate how long a time?
(A) There was some then.
(Q) There was what?
(A) There was some then.
(Q) There was some man?
(A) There was some t-h-e-n.
(Q) Some then. Oh, when you first see the object, there was some helicopters out—some helicopters in the sky? So, how many? How many are there in the sky when you first see the object?
(A) Three or four.
(Q) Three or four? Ok, you see three or four helicopters.
(A) And flying objects.
(Q) What kind of objects?
(A) Helicopters.
(Q) Helicopters. Are they the small ones or the big ones?
(A) Big ones.
(Q) Big ones, Ok, so you see three or four helicopters. Now you see some more coming in. Where do they come in? What direction do they …
(A) I don't know.
(Q) Don't know where they come in?
(A) I don't 'know when, but they were there.
(Q) They were there. OK, how about airplanes? Do you see any airplanes?
(A) No.
(Q) See any helicopters that look funny—that look different?
(A) Some wasn't as big as the others.

(Q) Some was bigger?
(A) Uh-huh.
(Q) Than the others?
(A) Uh-huh.
(Q) Some were bigger than the others?
(A) Uh-huh. One come close to the car.
(Q) One comes close. Is it hovering over the car?
(A) No.
(Q) Just comes close.
(A) Going around.

Vickie shifts to later in the time period when they had entered the car and began driving away:
(A) We stopped again.
(Q) You get in the car, and you drive, and you stop again?
(A) Uh-huh.
(Q) OK.
(A) And the helicopters was following it.
(Q) Helicopters following the object.
(A) Uh-huh.
(Q) About how far do you drive when you stop again?
(A) About five miles.
(Q) About five miles? And can you see the object some of the time?
(A) Uh-huh.
(Q) All of the time?
(A) Not all of the time. Sometimes the trees was in the way.
(Q) You can see it part of the time. And you can see the helicopters following the object?
(A) Uh-huh. All around it, too.
(Q) About the same number as before, about 20 ......
(A) Uh-huh. Yeah.
(Q) Can you estimate how close they get to the object?
(A) Uh-huh.
(Q) Hard to estimate?
(A) We could see.... over on 1960.
(Q) Over on where?
(A) When we come out on to 1960.

(Q) Oh, the road number?
(A) Yeah.
(Q) OK, when you stop again, what happens this time?
(A) They count 'em again. And I told her, I said don't be counting helicopters, let's get out of here.
(Q) Don't be counting helicopters? Betty gets out of the car again.
(A) No.
(Q) No? Just through the windshield?
(A) Uh-huh.
(Q) So, you didn't get out this time?
(A) That thing didn't disappear. It was still there.
(Q) It was still there? Ok, what else do you notice? Do you notice flame coming out just like before?
(A) No. After it got up there, it was just a long, red thing.
(Q) Just a long red thing. Yeah.
(A) It got far away then.
(Q) Far away, huh? So, it just looks like a long red thing in the sky?
(A) Uh-huh.
(Q) OK, what else do you notice? Do you notice the smell of lighter fluid?
(A) It's gone now.
(Q) It's gone now. Any sounds?
(A) Just—we can hear the helicopters.
(Q) You can hear the helicopters. How about any beeps? Can you hear any beeps?
(A) No.
(Q) How about any blue lights? Can you see any blue lights?
(A) No. 'Cause it's way away.

In seeking additional information about the incident, Dr. Sprinkle had Vickie picture the original encounter location and tell what else she could see. This line of questioning was aimed at determining whether or not she had seen any beings on the ground.

(Q) Picture yourself back the first time you were with Betty or Colby, you're in the car, and you're seeing the object. See yourself back in the car, see the object, and see the helicopters. Now, look around the road and see if you notice anything unusual around the road, anything happening on the ground.

(A) No.

(Q) Uh-huh. How about in the trees? Are there trees on either side of the road?

(A) Uh-huh.

(Q) Yeah. Notice anything about the trees that's unusual?

(A) No.

(Q) Any unusual light, reflected light, any motion in the trees? Did you notice any breeze or movement?

(A) No more than normal.

(Q) Normal breeze and normal wind. Yeah. OK, anything you can notice dropping from the sky?

(A) No.

Vickie told how they saw the object in the distance and how they "traveled and traveled and traveled" before they got close to it. She clearly described how close to the ground it came during their initial encounter and how that differed from the object's altitude the second time they stopped the car to watch it and the helicopters fly away. When the issue of heat came up, Vickie revealed the following details of her physical reaction to it. It all began when Vickie was describing how they still had the air conditioner on while they were stopped the second time.

(Q) You were looking up? Ok.

(A) It's hot.

(Q) Yeah, and is it hot?

(A) Not now!

(Q) Not when you are sitting in the car?

(A) Not when we—not when we stopped close to Super-Duper. It wasn't hot anymore. We had the air conditioner on.

(Q) You had the air conditioner on? And did you have the air conditioner on the first time the car stopped?

(A) We had to cut off the heater to turn the air conditioner on.

(Q) You had to turn off the heater and turn on the air conditioner?

(A) Right.

(Q) Yeah. And the second time you had...

(A) Air conditioner.

(Q) Air conditioner on. Because the car was so hot?

(A) Uh-huh.

(Q) OK, fine. How about your breathing? Is your breathing affected when the object is close? When it was spewing down the flames, was your breathing affected?

(A) Uh-huh. Uh-huh.

(Q) How does your breathing feel?

(A) Looks like I was going to choke to death, but I guess I was scared.

(Q) Was like you were choking when you experienced that? And it's partly because of the heat? Partly something else?

(A) 'Cause I was scared.

(Q) Anything else besides the heat and the fear?

(A) Smelled like lighter fluid.

(Q) And that lighter fluid, yeah. OK, do you want to say anything else about how you feel?

(A) Uh-huh, 'cept I'm burned.

Further questioning revealed that Vickie was quite firm in her belief that the U.S. Government was responsible for the incident. She told how she believed the object was man-made, not something natural like a meteorite and that the helicopters were real, government-owned craft also. She told how they could "hear the swish of the blades and the roar of the engines" as the helicopters went overhead. Throughout the hypnosis session, Vickie's emotions were quite clear—she was reliving a very real and traumatic event.

CHAPTER 24
The Pentagon Expresses An Interest

During the year following their encounter, Betty, Vickie, and Colby suffered greatly. They were frantic in their search for answers about what had happened to them. The U.S. Government became the focus of our investigation once we determined that no one but U.S. military units operated twin-rotor helicopters along the Gulf coast. During the next several months, letters were sent to the Department of the Air Force, the U.S. Army, U.S. Marines, Texas Air National Guard, U.S. Senators, and U.S. Representatives. We received no response to the letters, so they were followed up by telephone calls to everyone that received a letter and to all military bases within 300 miles of Houston. We didn't learn much.

This bureaucratic apathy was devastating to Betty and Vickie. After years of pride and trust in their government, they seemed to have become non-persons. As the year wore on, they became more frustrated and ready to cooperate with the print and broadcast media for assistance. Articles in Science Digest and Omni magazines were helpful, but it was only after the "That's Incredible" television segment aired that a Congressional inquiry forced the Pentagon to respond.

On February 24, 1982, I received a telephone call from U.S. Air Force Capt. Virginia A. Lampley. She worked for the Department of the Air Force, Office of the Secretary, Congressional Inquiry Division, Office of Legislative Liaison. She explained that a Congressional inquiry had resulted in her assignment to determine if USAF helicopters had been involved in the Cash-Landrum case. I replied with a thorough overview of the case, and she said she would get back to me.

About two weeks later, I heard from USAF Capt. Richard Niemtzow at Travis Air Force Base in California that Captain Lampley had concluded her investigation, and the results were negative. She found that the USAF doesn't use twin-rotor helicopters. I accepted that answer because some prominent aviation books said that CH-46's were used by the U.S. Navy, and the U.S. Marines and CH-47's were used by the U.S. Army. Several months later, however, I heard about a possible joint Army and Air Force operation involving the use of CH-47s. The August 17, 1982, issue of the Houston Chronicle newspaper showed a photograph of a twin-rotor helicopter. The caption read: "Honduran soldiers surround a U.S. Air Force helicopter during joint military exercise...." Unfortunately, this information came too late. Captain Lampley had already passed the case on the Department of the Army Office of the Inspector General in March.

On September 12, 1982, Larry W. Bryant, a long-time UFO researcher, sent me copies of notes and letters he had obtained from the Pentagon as a result of his Freedom of Information Act (FOIA) inquiries. In this material was a memorandum on Department of the Air Force, Office of the Secretary letterhead from Captain Lampley. It was a memo of record to the Army Legislative Liaison about her contact with me and was only partially censored. The uncensored part read as follows:

Mr. Schuessler has been in constant contact with individuals in the article. Severe medical problems were confirmed and ongoing. Mr. Schuessler approached an Army helicopter pilot with 136th Transportation Squadron, Ellington, who allegedly bragged about participation in the incident. Mr. Schuessler will provide the pilot's name and other info on request. Very interesting conversation with Mr. Schuessler."

I was called on March 19, 1982, by Lt. Col. George C. Sarran from the Department of the Army Inspector General's Office in the Pentagon. Colonel Saran explained that his office had received the inquiry from the Air Force Liaison Office because the Air Force concluded that their units were not involved. He explained further that his interest was in the possibility that Army helicopters were involved. He said he would be investigating that allegation. He stressed that the U.S. Army had no opinion about the unidentified object or UFOs in general.

Colonel Saran said he called me because his office had been requested to give some answers about the helicopter involvement. At his request, I provided a verbal account of the incident from start to finish.

Colonel Saran made it clear that he didn't believe the Army was involved. He stated that he had been stationed at Fort Hood in Texas before going to Washington and was familiar with their operations. For that reason, he felt that Fort Hood was probably not involved, although they have a number of twin-rotor helicopters. He said their testing and operations were generally conducted on the Fort Hood Reservation. Later in the investigation, however, he found that Fort Hood units fly all over Texas. In fact, in one of Colonel Saran's handwritten investigative notes released in -response to the FOIA request, he noted that on December 29, 1980, "700 *helicopters - Robert Grey field, came in, for effect.*" That note also said, "there may be other witnesses." Strangely, there was no official follow-up inquiry about these statements, and we didn't receive the FOIA information in time to force a more thorough investigation on behalf of the government.

Colonel Saran continued that initial telephone call by saying that as far as he knew, Fort Hood had the only helicopters of that type (i.e., CH-47 Chinook) in the area. In response, I told him I had found CH-47s to be stationed at Ellington Air Force Base in Houston and at the Dallas Naval Air Station and provided him with telephone numbers so he could check on them. He said that it would be his initial conclusion that if helicopters were present, they surely would have been from Ellington, and that would be the place for him to begin his investigation.

He had interpreted the incident to be a helicopter in trouble, landing for repairs, but concluded that didn't fit the situation because none had been reported. He had trouble accepting the Cash-Landrum claim that there was an object, probably a U.S. Government experiment other than a helicopter, in the air that night. He based his opinion on the fact this was the 1980 Christmas week, and most military installations go on holiday routine, allowing most of the troops to go home for the holiday period. He never went back later to reconcile this opinion with the fact that he found 100 helicopters landing at Robert Grey airfield on the same night.

After hearing about all of the other calls our investigators had made to various military installations, he concluded they were more or less truthful. He was very courteous and said he would try to contact Ellington. He said he'd be glad to act on any hint of "cover-up" that we might find, as the Army feels it very important to have a good rapport with the community.

Later the same day, Colonel Saran called a second time to let me know he had made contact with Ellington. The Commanding Officer of the 136th

Transportation Unit, a reserve group, stationed at Ellington, flying CH-47s, was Maj. Dennis Haire. He said he had instructed Major Haire to call and discuss the incident with me, although he had difficulty convincing Major Haire he was serious. He said it was obvious that Haire had never heard of the incident.

Armed with the information from Major Haire, Colonel Saran said now he was even more sure that no helicopter had gone down on December 29th. In addition, he assured me that the CH-47s were not flown on Monday nights. Since I had been monitoring CH-47 flights from Ellington periodically for several months, I immediately rejected his assertion by quoting from a log that listed three CH-47s flying out of Ellington on Monday, March 15, 1982, just four days before his call. It was obvious to me that someone was not telling the truth. Colonel Saran then agreed there were exceptions to the rule of no Monday night flights.

CHAPTER 25
Ellington Commander Responds

Tranquil setting: A sailor aboard the helicopter amphibious assault ship USS New Orleans finds a quiet spot at the stern to watch sea gulls cavort. The ship was docked at the San Diego Naval Station.

Our investigation revealed numerous media reports of helicopter operations from ships. This is counter to the assertions from the Air Force that this did not occur.

Major Haire called on March 22, 1982, as promised. He said he had been a member of the 136th Transportation Unit since 1966 and in charge since 1978. His detachment has eight CH-47A Chinooks, assigned there in 1980. Prior to that time, the group was a Medivac Unit.

He explained how the CH-47A has a fuel capacity for two hours of flying time plus a 15-minute reserve. Their cruising speed is 110 knots, with a

135-knot maximum. They can fly non-stop to San Antonio or Austin but must refuel at College Station if they fly to Dallas. If they go on a field exercise, they schedule a 5,000-gallon USAF fuel truck to meet them along the way. Each CH-47 takes 450 gallons of fuel for a fill-up.

Major Haire said there are no CH-47s in Louisiana. A contingent is stationed at Fort Sill, Oklahoma, and many CH-47s are stationed at Fort Hood, Texas. The Fort Hood CH-47s are the "C" model. They can do a round trip to Houston and back without refueling. They have a three-hour-plus 30-minute reserve capacity. Their maximum speed is 175 knots, and their cruising speed is 140 knots. He verified that there were no Chinooks flying in 1980 except for military units, so civilian Chinooks could not have been involved in the Cash-Landrum incident.

The Ellington unit flies around the Houston area all of the time. They average 2,000 hours per year per pilot. Fort Hood averages 900 hours per year per pilot. Major Haire is proud of his unit's record. In addition to the eight CH-47s, they have four Hueys and four '58s. The unit does a lot of airborne troop emplacement drilling with these aircraft. They use the Addicks Reservoir north of Houston as a jump zone. To get there from their Ellington Field base southeast of Houston, they stay below the 1,800 feet minimum for the Houston Intercontinental Airport Terminal Control Area (TCA) approach paths. This is the same modus operandi used by the flight of helicopters involved in the Cash-Landrum encounter. However, Major Haire said he was 99% sure the Ellington CH-47s were not involved. He said this could be verified by reviewing flight plans, aviator's records, and aircraft records which we later requested in the discovery part of the legal case against the U.S. Government. The official response was that the records had been destroyed.

According to Major Haire, each flight of a CH-47 requires two pilots plus an enlisted crew chief. Sometimes a fourth person joins the crew. The CH-47 can pick up and carry small equipment, but nothing very large. He said that large loads would require a Sikorsky Flying Crane. I was not able to pin down what he meant by "very large" because I have a variety of photographs of CH-47s carrying an automobile, an Army howitzer, and an F-4 fighter aircraft.

Major Haire also said helicopter pilots are very light-sensitive at night and try to avoid bright objects because they ruin the pilot's night vision. They don't even turn on the inside helicopter lights until after they have landed. For that reason, he doubts that helicopters would have flown near the diamond-shaped bright object described by Betty and Vickie.

In addition to being the Commanding Officer of the 136th Transportation Unit, Haire operated a commercial helicopter service at Lakeside Airport in Houston. At a later date, Major Haire was hired by a documentary film crew to fly a simulation of the UFO encounter at the scene of the original incident. His excellent piloting skills were evident when he was able to repeatedly land at night in among the trees, scrub brush, and weeds beside the road. Car lights didn't seem to bother him as he zipped in and out of the path cut between the trees by FM1485 as he buzzed Betty's car in the simulation of the original event.

The Major called me again on March 26 to tell me that he had shared the results of our telephone conversation with Colonel Sarran. He said Colonel Sarran is digging into the case like a tiger, representing the Army Inspector General's office. Major Haire again said he had no idea what was going on near Huffman that December night, but he was definitely not involved. He said he felt that the government well might have some special devices, some advanced technology, or some test vehicle that could cause the reported symptoms. However, as a civilian or as a National Guard member, he doesn't know of anything like that.

I ask him if he had ever heard of a NEST unit operating here. He said no, and he didn't even know what the acronym meant. I explained that it meant Nuclear Emergency Survival Team, an elite group ready to help recovery from terrorist or other nuclear accidents. He drew a blank on it.

CHAPTER 26
A Helicopter Comes To Dayton

April 30, 1981 was a busy day in the small town of Dayton, Texas. The Future Farmers of America was holding a livestock show, and the whole town was buzzing with activity. One of the special events was the static display of a CH-47 helicopter on the big vacant lot where the Savings and Loan company was built soon after the event.

When the helicopter flew over Dayton in preparation for landing, Vickie saw it and called her friend Martha Thompson. Vickie said: "Martha, for God's sake, get over here. One of those helicopters that was with that thing is flying over the house right now, and I'm going to follow it."

Martha said: "Vickie, I'll be there in just a second."

while she was waiting, Vickie called me and asked what to do. My response was to ask questions and get the pilot's name. Vickie was determined to get to the bottom of the mystery now that one of the helicopters was so nearby. Then Martha picked up Vickie and Colby, and they went to see the helicopter.

Sometime later, Bob Pratt and I were at Vickie's home getting more information about the helicopter visit to Dayton when Martha and her husband Larry came over. During the interview, Vickie, Colby, and Martha told us that it was a clear, bright, warm day, and they had to stand in line to enter the helicopter. They were both excited and frightened at the same time by the experience. The transcript of that interview is as follows:

(Q) Martha, you were with Vickie and Colby the day of the livestock show when a helicopter was there, and you and Vickie talked to a pilot. Can you recall that conversation?

(Martha) Yes, sir.

(Q) Can you recall what was said between the two of them?

At a Livestock Show in Dayton, TX, Colby is able to get close to a CH-47. April 30, 1981

(MARTHA) Yes. Vickie commented to him that she has seen a whole slew of helicopters just like this one the night of her encounter, and he made the remark, "Yes ma'am, we were all called out that night." We had taken my two little kids up there and we acted like we wanted his autograph for one of them. And so that's when he wrote down his name and everything, and he said his platoon was called out that night and some other ones...

(Q) Called out for what purpose, did you say?

(VICKIE) Anything, any purpose other than just show. And he said:

"Yes, ma'am."

And that's when he said they were called out on that night by the Montgomery County Sheriff's Department.

(Colby interjected) I sat in the pilot's seat.

(Q) You (Colby) mentioned something to him about being hurt by this thing? Did you or your grandma say something about being hurt by this UFO?

(COLBY) Yes.

(VICKIE) And that's when he clammed up.

(Martha) Backed off.

(Q) You were with Vickie at this time?

(MARTHA) Yes, sir.

(Q) You did get the autograph?

(MARTHA) Uh-huh.

(Q) He said other units were called out, too. Were there any comments of any kind that gave you an idea of what these other units were?

(MARTHA) You mean the different units that were called out?

(Q) Yeah.

(MARTHA) The way, if I'm not mistaken, what the man said was that his unit was called out and a "bunch of the others."

(Q) Did you try to ask him any questions about this?

(MARTHA) Yeah. Vickie conversed with him quite a bit.

(VICKIE) That's when he clammed up.

(MARTHA) I didn't. I knew she (Vickie) was excited, because evidently this was the first time, she had seen one of those since that night, and I didn't know exactly what kind of questions she wanted to ask him, except that we had already arranged before we went up there to try to get the pilot's name if we possibly could. And I said: "Well, shoot fire, we'll just say one of the kids wants it." And that's really the only thing we agreed on before we went into it, was to try to get his autograph. So, Vickie really did most of the talking, but I did hear the gentleman say they were called out on that night. Vickie mentioned the date to him and everything, and he said: "Oh, yes, we were called out—our unit was called out that night."

(VICKIE) They were called out, he said, around 9 o'clock that night. He thought about 9:15 or 9:20.

(MARTHA) And then he made the remark about it being by the Montgomery County Sheriff's Department. And Vickie repeated it. You know, after he said that, Vickie said: "By the Montgomery Sheriff?" and he said, "Yes."

(VICKIE) I said Glory be to hallelujah, I found somebody that knows what we run into. I was talking to Martha and Colby. He clammed up just like that.

(MARTHA) That was a mistake right there.

(Q) He wouldn't talk about it anymore or what?

(VICKIE) No.

(Q) Did he walk away from you then or what?

(MARTHA) Yeah.

(Vickie) It looked like he was embarrassed and said something that he shouldn't say. That's the way he acted.

(Q) You say he was embarrassed?

(MARTHA) That is the way he was acting. That, and then they had people filing through this thing looking at it (the helicopter), and we were up in the front of it. We had people behind us, and it was terribly hot.

(Q) Were you inside or outside of it?

(MARTHA) We were inside, and it was a scorcher that day. Sweat was just pouring off of everybody, especially the guys (crew members) because they were in their fatigues, and he just acted like he kind of wanted to push us off. Now I went outside with Colby. If I remember correctly, Vickie, you stayed in there a few minutes more.

(VICKIE) Yeah, because I was trying to get every bit of information I could.

(MARTHA) She was trying to get more out of him, and he wouldn't say nothing. He kind of gave me the impression that you know—the impression that he was trying to imply that: "Well, I've got people back here, and you're holding up the line." So, I took Colby and went on out, and Vickie just stepped over and stayed there with him.

(Q) What kind of markings did he have on his uniform? Did he have any bars, any silver bars or anything?

(MARTHA) I'm sorry I just can't remember. They had on those green, real heavy khaki outfits with big black high-top boots.

(Q) Did he have a name tag on?

(MARTHA) Uh-huh. It had U.S. Army on the side of the helicopter. Vickie took pictures of it that day.

(VICKIE) I went back home and got a camera and film, and we went back up there, and I shot it. It was a little bitty camera, but I got 'em.

(Q) Were there any other members of the crew nearby when you were talking to him?

(MARTHA) No, nobody else.

(Q) Where were they?

(MARTHA) More to the back.

(VICKIE) Others didn't say much, period. They were the ones that were waiting to march in the parade.

(MARTHA) Now, I never heard any conversation with anybody else except this black gentleman. He's the only one that I heard say anything about that night, about the helicopters being called out or anything else.

(Q) And he is the one that signed his autograph and said "CH-47 pilot?" (Signed: Willy R. Culberson, CH-47 Pilot).

(MARTHA) Yes sir.

(VICKIE) Yes sir.

(Q) OK, who got the autograph, actually?

(MARTHA) My son wanted an autograph. We got it for him.

(Q) OK, you actually got it for your boy.

(MARTHA) Well, yeah, I guess more or less, 'cause what was it we wrote it on, Vickie?

(VICKIE) Envelope.

(Q) Where was the envelope, in your purse or where?

(MARTHA) No, we got that out of the glove box of the car, and we tore off a piece.

(Q) Was this before or after you had talked about the UFO?

(MARTHA) We got his autograph before the gentleman started backing off, before we actually started questioning him. We got the autograph prior to that. He had already made the statement, though, that his unit was called out that night because Vickie had said ~ that was one of the very first things she said to him, "I seen a bunch of these the night of uh December 29th." And he said, "Oh, yes ma'am," and made his comment. And I told him my son would like to have the autograph of the pilot. I asked him, I said, "Sir, are you the pilot of this aircraft?" And he said, "Yes, I am." And I said, "Well, my son would like to have your autograph."

(Q) You (Vickie) stayed in the helicopter for a while after she left?

(VICKIE) Uh-huh.

(Q) Did you continue trying to talk to him at that time?

(VICKIE) Yes. He just evaded it. He wouldn't say nothing. He'd done fouled up and he didn't know how to get out of it. See, Colby was with me, and he had them blisters on his face and he wasn't able to go to school and I told him, this is the little boy that was there that night. And I says, and I'm the woman. And my eyes was just — I said, see my eyes? And he didn't say not one word, not one word.

(MARTHA) And Colby was so scared. 'Cause when Vickie called and told me to bring the car over to her house, you could hear Colby in the background just screaming bloody murder. "No, Grandma! I don't want to go see it!" And he was just screaming, and Vickie had to tell him to be quiet so she could hear me talk. I remember hearing him screaming. And then when he seen all the other men and women and kids going through it (the helicopter), we convinced him that it would be safe for him to go inside and see.

CHAPTER 27
The Helicopter Pilot

Finding the helicopter pilot gave Betty and Vickie renewed hope that the U.S. Government would give their doctors information about the device that had harmed them so that the doctors could give them life-saving help. It was the goal we had all been striving for.

Believing this was a piece of evidence equivalent to a "smoking gun" in criminal investigation, I spoke directly with Officer Culbertson, relayed the information to the U.S. Army Inspector General's Office, and discussed it with his commanding officer, Major Haire. I was completely open and frank with each of them. This was a chance to help the victims, and I was asking for their participation.

In my discussions with Culberson, I explained that I worked for a contractor at the NASA Johnson Space Center. In addition, I represented a non-profit UFO investigative organization. I told him how I had been alerted to the case by the NASA Public Affairs Office because of my personal interest in UFOs and that NASA was not in that business. He laughed lightly in response.

When I told him about the large number of helicopters involved in the incident, he said: "Uh-huh." And when I said there was an unidentified object involved, he said: "yes."

I told him that "I had word that maybe he was one of the pilots at the scene at the time," he said: "Uh, no I wasn't, and I'm familiar with that call because I talked with that lady, that lady who reported the same thing; but we were not involved with that at all." I said: "Is that right?" And he answered: "No."

I explained how the two ladies and little boy were burned so badly and how we were seeking information that would help treat them but were not having much luck.

He explained, "The only reason I'm familiar with it is I took a CH-47 to Dayton, Texas. And the lady came up, and she reported the same thing to me. It was her, and I think she said someone related to her, and the little boy was with her. And her face was burned, and her hair was falling out, and it was affecting her eyes."

I asked: "You could actually see that?" He said: "Yes."

He said he tried to find out for himself about the aircraft she said was exactly like the one they had there in Dayton, a CH-47. He said: "But the only people in this area who have CH-47s, uh, we're the only ones who have 'em, and the next closest CH-47 is the regular Army at Fort Hood."

We talked about the Fort Hood officer saying they had a large number of helicopters up that night but denying they were in the Houston area. We also discussed the CH-47s located at Dallas, but he said that if anything was going on in the Lake Houston area that his unit, not the Dallas unit would be involved. To that, I said, "None of your people happened to be doing that, though, huh?" He replied: "No."

On March 26, 1982, I spoke to Major Haire about Culberson's potential involvement, with the following results: Major Haire knows Culberson well. In fact, he had seen him just one day earlier when the CH-47s last flew. He said Culberson is not a part-timer like himself. He is in charge of maintenance—a full-time position. When the Guard unit flies, he participates with them. He is a highly qualified, top-notch officer. Haire verified that Culberson flew the CH-47 that was on display in Dayton. In this case, it was a weekday flight, and Culberson, being a full-time person, was the natural one to fly the mission. It is difficult to find reservists that can fly during the week from 8 a.m. to 5 p.m. because they have other jobs.

Major Haire made an interesting point. He said their unit could not, by law, respond to an emergency called by a sheriff or any other civilian. Only the Army Adjutant General, Willy Scott, in Austin, could authorize either an official or non-official response to such a call. Even a call from Maj. Del Livingston in Dallas would not get a response. Major Haire did not comment on whether or not Willy Scott had authorized any flights during the time period in question; therefore, I assume the answer was negative. He said that the call-up rule did not apply to an Army Reserve unit like the one located at Tomball, Texas, just north of Houston, so we looked into that unit's activities also.

As a part of the Army Inspector General's investigation, Colonel Sarran talked to Willie Culberson at Ellington Field. He said Willie "thought the Montgomery County Sheriff Department called someone." Sarran said, "he was just talking." Personally, he was not flying that night. Colonel Sarran said his report will list his investigation as "inconclusive."

CHAPTER 28
The Army's Investigation

Colonel Sarran called again on April 8, 1982, to arrange for a trip to Houston. He said the investigation by the Office of the Army Inspector General was underway. He had already talked with U.S. Air Force Capt. Richard Niemtzow at Travis AFB and with Dr. Peter Rank in Madison, Wisconsin. He understood their viewpoints but decided to come to Houston in May to investigate for himself. He said there was nothing secret about his involvement. It was an open investigation. He said: "The Army doesn't say UFOs exist or do not exist. That is up to someone else. If Army helicopters were involved and it was their fault, then the Army would take responsibility."

He went on to say that all reserve units are under Force Command in Atlanta, but this doesn't apply to National Guard units. Willy Culberson is in the Guard. He is an Assistant Staff Technician (AST). Later, on May 11, Colonel Sarran told me that Culberson was a Chief Warrant Officer (CW3). This was enlightening because as a result of my calls to Ellington, I had been led to believe he was a Major. The revelation was a surprise to both of us.

Colonel Sarran said he checked to see if any helicopters sprayed fuel or were involved in an agent orange-type drill. He received negative replies from the Training Indoctrination Command, the Testing Agency at Fort Hood, the Corpus Christi Naval Air Station, the Aberdeen Proving Ground in Maryland, and the Pentagon R7D. The computer lists no activity at Huffman, Texas, on December 29, 1980. From what we found out later about the Helicopter Amphibious Assault Ship operations, perhaps the computer search should have been broader than just "Huffman, Texas."

He called again on April 23 to discuss his progress to date and to set a date in late May for his trip to Houston to interview me, Vickie, Willie Culberson, and a Dayton policeman who had also seen the helicopters.

He said he had contacted all bases and found that none flew near Huffman, Texas, on the subject date. Fort Hood was now saying they flew only one helicopter on that date, not 100 as previously reported, and it flew to Houston, on to Galveston, and back to Fort Hood, arriving before 8 p.m. He said he had exhausted all resources available to him.

On May 25, I met with Colonel Sarran at the Houston International Airport Holiday Inn. At that time, he requested that I do a taped interview relating what I knew about the case. I agreed to the interview as long as I could also tape the whole thing. He cordially agreed, and I proceeded to give him a fairly detailed picture of what had transpired up to this point in time.

Later in the day, I took Colonel Sarran to Vickie's house in Dayton so he could continue his investigation with her. Vickie's sister Bertha was also present throughout the interview.

Colonel Sarran told Vickie this was an official investigation, but she was free to talk about it to anyone. He then gave her a form letter under the Privacy Act which he said was designed to protect her. However, he told her he could not guarantee privacy.

The next step in his investigation was an interview with Betty. He called Betty in Birmingham from Vickie's house and went through the same interview routine he had used on Vickie. He said he was satisfied with the interviews and found Betty and Vickie to be open and forthright.

From Vickie's home, we went to the home of Dayton Police Officer L.L. Walker, another witness to the helicopter flights. That interview is covered in detail in the next chapter.

We also visited the offices of the Montgomery County Sheriff's Department in Conroe, Texas, to check on the claim that the sheriff had instigated the flight of helicopters that flew around the UFO that night in December. What we found was as strange as the UFO encounter itself. We were told that there had been an administration change as of January 1, 1981, and no one working in that office on December 29 was still employed there. In fact, we were told that none of the new employees knew any of the old employee's names or where they lived. I asked what kind of police agency lacks the skills to look in the files and find out who was there a few weeks earlier, but that didn't help.

We spoke with Chief Deputy B.J. Grounds, Lieutenant Lowre, and Pete Perkins. None of the people interviewed said they would call the National Guard. It was not a part of their normal procedure. Instead, they would probably call the Houston Police Department if they needed officer assistance.

Interestingly, Chief Grounds suggested we contact Carl Mangogna, the Harris County Patrol Division officer in charge of helicopter flights at the time of the incident. We tried to contact Mangogna at the Harris County Sheriff's Department and found he was no longer employed there. They did tell us that Mangogna was currently employed as Chief of Security for the Kaneb Corporation of Houston. We went to the Kaneb offices but were unable to contact Mangogna. The next day I tracked Mangogna down by contacting his father, also named Carl Mangogna. After I explained the thrust of our investigation, Mangogna said he "had no memory of the events on December 29, 1980." However, he did suggest we call Captain DeFore of the Houston Police Department.

Colonel Sarran spoke with Captain DeFore about the incident. DeFore had no knowledge of what had happened. He did agree to go see Officer Walker and talk to him about his report of seeing the helicopters.

On May 27, Captain DeFore reported that he had no doubt that Officer Walker had seen CH-47s. DeFore said he felt they were probably part of a Quick React Force. He reiterated Walker's report of 12 helicopters flying in groups of three and shining spotlights on the ground as they flew at low altitude. Another group flew about one and one-half miles behind the others. All were no higher than 500 feet in altitude. He said they could have been from anywhere and refueled from any of the 5,000-gallon fuel bladders in the area, or from a flat-top (ship) in the Gulf of Mexico.

Chief Grounds of the Montgomery County Sheriff's Department suggested we carry our investigation to the Army Medivac Unit at Hooks Airport located to the northwest of Houston. We went there but found the place closed. All the members of the unit were at a special meeting out of town.

On May 26, Colonel Sarran was able to speak with Chief Warrant Officer W.S. Gustafson, head of the Army Medivac Unit. Sarran was impressed with Gustafson. Gustafson, he said, was a very sharp individual and would be helpful in the investigation. Sarran told Gustafson to call me if anything surfaced.

A few hours later, Gustafson called me with his report. He had been busy. He said he had gone back through the Hooks Airport records and found the weather data for December 29. The wind was 210-230 degrees at 10-15 miles per hour, with no rain, with the temperature in the mid-'40's. The date of the full moon was December 26.

His discussions with other members of the Medivac unit led to the following initial conclusions:

1) There was a Quick React Force operating in Louisiana and Texas periodically for the past year and a half. They last heard of it about six months earlier, operating near Morgan. City, Louisiana. He said they were practicing "Iran-type" raids from a small carrier in the Gulf of Mexico. At times, he said, they haul in and strategically place 5,000-gallon fuel bladders for refueling the helicopters during an operation. He said their operations are secret and are not announced.

2) The USMC in New Orleans operates CH-46 helicopters, similar in appearance to the CH-47s. He was not able to find out if any were in the Houston area on the date in question.

3) He suggested a follow-up with HPD Captain Ken DeFore and have him speak with Officer L.L. Walker. Coincidently, DeFore also lived in DA.

I passed all this information on to Colonel Sarran for verification through his official channels. However, Sarran told me he was having no luck in finding any related information in the government channels. As a result, we discussed who else might be able to help him. I suggested several individuals, but the most likely one was another Army officer, Col. John Alexander. Sarran said he would give Alexander a try. In a later conversation, he told me he was never able to make the connection with Alexander.

The next day, May 27th, Chief Gustafson called again with more information. He said Captain DeFore had talked with L.L. Walker and was convinced Walker had seen a large number of CH-47s. DeFore also believed they were a pan of a Quick React Force. After describing DeFore's description of the CH-47 flight activities, Gustafson said: "We may have uncovered a bucket of worms." This information was also relayed back to Sarran.

Colonel Sarran called again on June 1. He said he had no luck locating a unit responsible for the helicopters. He had checked to see if it was a Quick

React unit out of Fort Benning, Georgia, but that is not a Quick React base. Only Fort Bragg, N.C., Fort Devons, MA, Panama, and Europe have Quick React Units, and none were involved. There are no Quick React units located on ships.

With the help of a number of people, I had been keeping track of a lot of multiple-aircraft helicopter flights in and around Houston. The latest was a flight of five CH-47s over Dayton on May 22nd at 11:00 a.m. I asked Sarran to check and see where they were from and where they were going because that might lead us back to the origin of the earlier flights. He said it would not help. He said lots of units could overfly the area. Fort Hood participates in the yearly Reforger exercise and flies from Fort Hood to Port Arthur, Texas, as a part of the exercise. That would take them near Dayton. Vast numbers of helicopters are involved. I was somewhat agitated by this answer since I had been told by a number of military people that Dayton was too far from any helicopter base for operations to take place there. However, he was undaunted. He said the computers listed no operations at Huffman, and there was nothing else he could do. He was due to answer the original inquiry and talk to John Nyter, Deputy Head of Congressional Liaison. He said he would get back to me on "the Quick React thing" if he found anything, but he expected "negative findings."

Colonel Sarran's last call was on June 25. The investigation had gone nowhere. He said he had contacted the lawyers at Bergstrom Air Force Base about their discussions with Betty, Vickie, and Colby. He verified they were there at the suggestion of Senator Tower and Senator Bentsen. His review of the tape made during the interviews revealed nothing new. He did say: "Obviously, something happened to the ladies;" however, he could find no group responsible for the helicopters. An operation of that magnitude would have been "big time." Fueling would have been a problem. And he just couldn't believe it wouldn't have been exposed. He said the Special Operations Branch, Delta Project, and the skyjack and terrorists fighting groups all responded with a "negative" 10 his inquiries about their possible involvement.

Coral Lorenzen of the Aerial Phenomena Research Organization claimed knowledge of a flight of an experimental craft from Albuquerque, New Mexico, to Ellington Field in Texas, on December 29. She felt this explained the Cash-Landrum encounter. Colonel Sarran called Coral and questioned her about her information. He concluded she had nothing new to add.

CHAPTER 29
Police Officer Saw Helicopters

On May 25th, Colonel Sarran and I, accompanied by Vickie Landrum, met with Dayton Police Officer L.L. Walker and his wife Marie in their home in Dayton, Texas. Colonel Sarran continued his investigation into the allegation of military helicopter activity during the Cash-Landrum encounter. The interview proceeded as follows:

(SARRAN) OK, just give me your name and where you work, and where you live, and then, I'll just start asking some questions as to what might have happened that day. OK, this is pursuant to an inquiry that we're doing in regard to a phenomenon that happened back in December of 1980 in which this gentleman and his wife have knowledge and can testify as to what happened. And they are both here today on the 25th of May at approximately 1:30 p.m. in the afternoon. So, at this time, would you state your name and where you work, and then we'll talk about what might have happened that day.

(LAMAR) My name is Lamar L. Walker commonly known as L.L. Walker. I've been employed by the City of Dayton since 1964. I am a patrolman in this city and also a certified state police officer. On the date in question my wife and I...

(SARRAN) Now which date was that 'do you recall what date it was?

(LAMAR) On December 29th my wife and I was coming back home from her mother and Dad's home who live in Plum Grove. It's about three miles behind Splendora into the wood area. We was traveling New Caney Road. We just came through there, the cut-off, and hit Cedar Bayou and came across the river and cut down the school road [Huffman-Eastgate

Road] at the Huffman new High School there and just got back on FM 1960. We was inside the Liberty County city limits and just made a turn out there by the railroad tracks on a curve, headed east.

And I made a remark--! said, "Marie," I said. She said, "What's that noise? I said, "Well, I don't know." "But," I said, "It sounds like helicopters, and it is getting louder." She said, "Well, I don't see any airplane." And I said: "It's not an airplane—it's a helicopter, Marie." And she said, "Whatever it is, it sure is low." And I said, "Yeah, it is." So, I rolled my car window down, and there was very, very little traffic. So, I slowed way down, and I started looking, and I could see some flashing lights in the air approximately anywhere from 400 to 500 feet in the air.

And I got to picking out more of them, and as I was picking them out, I picked out three in a victor (V) formation and about a thousand [feet away]. And a little bit off to the left of it was another sector of V-with three choppers in it. And as I looked a little bit better, I seen three more. I know it was three sets for certain. The twin props, front and aft, the shape and everything. I said, well, they must be on maneuvers again—National Guard or something, out of Fort Polk or the Coast Guard doing something. And I looked a little bit closer, and you could see some lower lights back off in the distance quite a ways back. I'd say about 3/4 of a mile. The visibility was real good that night.

(SARRAN) When you saw the helicopters. How many? You say it was rows of Vs.

(LAMAR) It was victor of threes in each victor.

(MARIE) 'Bout 8 or 9 of 'em what it looked like to me.

(SARRAN) 8 or 9 rows, with three, three per row, is that it?

(MARIE) Well, they weren't all together. They were separated going across the road. They had real bright lights on 'em like they were searching, looking for something.

(SARRAN) Uh-huh. Did you see any kind of real bright phenomena? Anything that was brighter than anything else? You say they had landing lights on?

(LAMAR) They possibly had a landing light or some kind of searchlight on top of 'em. First lead two had some type of a light shining down.

(SARRAN) Did it have twin rotors on it or just one rotor?

(LAMAR) They had twin, forward and aft.

(SARRAN) Forward and aft?

(LAMAR) Yes.

(SARRAN) Did you see any other helicopters beside the ones that had the two... Did you see one with just one rotor on it? Or did all of them have twin rotors?

(LAMAR) I can't recall seeing any with one rotor on it. 'Course, like I said the ones that I did see, and I'd recognize, uh, heck they wasn't off the highway, oh heck, it was close enough just their running lights and everything and enough starlight and everything and moon and everything that I could tell what they were by the outlines and everything.

And I told Marie, I said: "I don't know what they're looking for, but the way they got those searchlights running down and everything-they were bound to have an airplane down somewhere." Because they was flying a lead ship here and one about maybe, oh as much as maybe a hundred yards, and a little bit back here, and one here and then over here there's another three set up, and then back down behind 'em there was some coming back down there, flying like they was separated for some reason or another, I don't know what.

(SARRAN) You remember what day of the week this was?

(LAMAR) No I don't-well, I tell you what, it had to be one night when I wasn't working. See, I'm off Monday and Tuesdays. [It took a fair amount of discussion for the Walkers to work through defining the exact night, but they did it two ways. First, they pinned it down to the start of the week that Betty went into the hospital. Second, Lamar had worked at the bank on Monday, and they had gone to her parents' house after he got off work. They were able to pin the date down to December 29].

(SARRAN) And how far from this location here [pointing to the sighting location on a map] did you see the helicopters? I mean did you see them ten miles to the south.

(LAMAR) No, you couldn't see that far—lemmie have a scratch pad-OK, this is FM 1960. The railroad tracks are here out on 1960 approximately eight miles out, and it's about another, I'd say a good mile, maybe a mile and a quarter there's a side road that cuts in [Huffman-Eastgate]. Here's your new Crosby High School here and that road is approximately...

(VICKIE) Uh, Huffman...

(LAMAR) Huffman—! keep calling it Crosby. Let's see that road, I'd say is a good maybe 3-4 miles [FM 2100] and it cuts into another road that takes you into the New Caney [FM 1485] area back through the woods, and back in here is when we first saw...

(MARIE) Right in here is where we saw 'em [pointing to the area near where Betty stopped the car for a second time to watch the helicopters fly across and out of the area].

(LAMAR) We saw 'em right in here and they was coming out in a direction. We saw 'em coming out of the direction something like this [east] and they made a turn right here [going southwest] and I guess about the closest part there is about 3/4 of a mile. We watched them all the way from this point back across this railroad track and after they done got by, we saw these other widely spaced helicopters running lights behind them.

(SARRAN) And they were military helicopters?

(LAMAR) Definitely for sure. Was definitely sure military helicopters [Lamar said he had flown in the same type aircraft].

(SARRAN) Did anybody else see the helicopters that you know of?

(LAMAR) Now let me think. I talked to one officer I believe that said he had heard some, but he never did see 'em. He said that they passed... oh, wait a minute, by golly...

(MARIE) There was a pick-up that passed us.

(LAMAR) No, no, no. Come to think of it old major, uh, what's his wife's name and his name? She works for the bank and in bookkeeping. He's an ex-major of the army helicopter section and served in Vietnam. He was over at his house when they come by and he happened to go out and look at 'em. And he was wondering what happened 'cause his wife asked me the next morning at the bank.

(SCHUESSLER) You have been saying the name of the helicopters is Huey, but you say they have two rotors.

(LAMAR) Two rotors. It's a big Huey copter. I may be mispronouncing the name of the vehicle. It has two blades. In fact, it's loaded by the ramp in the back. It opens up and you can put a jeep in there, or use it as paratroopers, or just whatever you want to.

(SCHUESSLER) Is it possible you are talking about a Chinook?

(LAMAR) It might be the Chinook. It's a large type ......

(SCHUESSLER) But it is a twin rotor?

(LAMAR) Right, it is a twin rotor.

(SARRAN) Yeah, you are talking about a Chinook, not a Huey. Huey only has one rotor.

(LAMAR) Well, that's where I stand corrected. It was a big workhorse of a...

(SARRAN) But that's all you saw. You didn't see any one rotor helicopters?

(LAMAR) No, not to my estimation I didn't.

(SARRAN) And they were in groups of three?

(LAMAR) I do remember the groups of three and I couldn't understand why they were flying in that formation.

(Saran) Can you tell me roughly over a period of, what, 5 minutes, how many did you see, in groups of three?

(LAMAR) They wasn't exactly in a line abreast. They were more or less staggered one here and one here and one back here following them [pointing to the location on the map]. The others had done got outa sight, but we could still hear 'em when we saw the other three.

(SARRAN) So, how many, 10, 12?

(LAMAR) I'd say all together we saw 3, 6, 9--I'd say 12 altogether.

Col. Sarran and I cross-questioned Officer Lamar at length going over the material a number of times. His answers were all consistent and apparently honest. In fact, at one point he described the helicopter landed in Dayton by Willie Culberson as being exactly the type he saw that night. Then we went back and reverified his statements about how he was driving at the time of the sighting, as follows:

(SCHUESSLER) And you went how far from the time you first saw them until you didn't see them anymore?

(LAMAR) We saw around this area [pointing to the map], and I'd say it's about a mile and three quarters to two miles to the railroad tracks. That's where I lost contact. You got a curve here and I saw some headlights coming at me.

(SARRAN) So you were going in this direction [east] and the helicopters were going in that direction [west].

(MARIE) We were about right here when they first came across. And then... some more came.

(LAMAR) They were headed in this direction [east] and they kind of made a small vector like they was going to cross the road up over in here somewhere. I don't know which way they were going after that.

(SCHUESSLER) You had them in sight for about a mile then. You said you slowed down. How slow did you get?

(LAMAR) In fact I first started hearing the noise and I first thought something was wrong with the car...

(MARIE) We stopped.

(LAMAR) Oh, very slow. Yeah, I did stop and got off the side of the road and watched them a little bit.

(SCHUESSLER) So you didn't zip past at 60 mph...

(LAMAR) Oh no, I observed 'em and got a real good look. In fact, with the running lights and the spotlights they had behind them, the way was clear enough you can see the outline of 'em real good. In fact, I was wondering what they was doing that low.

CHAPTER 30
Other Reports Of Helicopters

Numerous individuals came forward to tell about seeing large numbers of helicopters the same night. Some were as close as a mile from the incident site, while others were as much as 10-12 miles away. For example, a man living on County Line Road 4, to the east near the Trinity River, said he saw twin-rotor helicopters fly over his house in "droves." An oil company executive living in Crosby, to the south, heard the helicopter noise and rushed outside to observe a large number of helicopters fly overhead at a very low altitude. They were so low the house seemed to vibrate in resonance with their rotor blades. A man was outside his home a short distance to the west of the site, playing with a Christmas gift toy with his 10-year-old son when five twin-rotor helicopters flew overhead. He said they were going southeast and were only about 800 feet up. He was positive they were CH-47s because he was in the Army and exposed to CH-47s on a regular basis. He said he remarked at the time, "the army has something going on tonight."

Two very good witnesses, also living in Crosby, were Rosalie Seymour and her daughter Michelle. They saw the helicopters at very close range and at the same time, saw the bright object near Huffman while standing on the lawn of their home. When they heard of Vickie and Colby's plight, they hoped to be able to help by describing their observations. They even offered to appear in court if necessary. Mrs. Semour's husband Marvin was at work during the incident but was told all about it as soon as he arrived home that night. During the interviews, he offered some interesting observations about the helicopters.

Excerpts from their taped interviews follow:

(ROSALIE) I was in here in the house. Michelle was in her room hooked up listening to her stereo like she always does, and she heard this racket. And she came to here, and she said, "You hear that racket?" And I said, "No." So, she went on outside, and in a few minutes, the door opened, and she yelled, "Mama, come here and look." And I went outside, and I looked over here and [it] crossed over Lucy Bell's [house]. I saw this helicopter coming. And then over toward Alice's was another helicopter coming. And right at first, I saw what I saw, and I said, my God, there must have been a wreck on [Hwy] 90, and that's the Life Flite [hospital helicopter]. But it wasn't a Life Flite. So, I got to looking, and everywhere I looked the sky was full of helicopters. They were coming from every direction, like from Houston and this direction [pointing north]. The sky, they were low—but it was just full of helicopters.

First of all, we went out here under the carport, and over there [pointing toward Huffman] was this big glow just over the top of the tree line. You could see this glow. And I said: "Well, undoubtedly, it's been a train wreck." And then we got to thinking, where in the hell would the tracks be. The tracks are right here. How could there be a train wreck over there? So then, I decided it must be a forest fire; but I never could see any flames. But I could see the whole sky was lit up like it was coming from the ground. And it was just a glow-like. There wasn't no flames or nothing. I remember telling my husband about it when he came in.

(MARVIN) I was out there at work, and they had a newspaper article in the *POST* or in the *CHRONICLE,* and I just tore it out of there, and I just brought it home, and I just come in that night 'cause I knew what it was. I knew it was what she had seen. So, I just brought the newspaper in, and I said, "Look here, Rose, read this." She read it and said, "That's the same night, y'know, that's the same night." So, we looked it up on the calendar there [pointing to a calendar on the wall]. I had a calendar where I had worked overtime. And that particular week there, that was the only night I worked that week. So, we went in there and looked, and sure enough, that was the date, December 29th., and sure enough that was it.

(ROSALIE) I did see the helicopters, and so did my daughter.

(Q) Michelle, about what time did you say it was?

(MICHELLE) About 9, around in there.

(Q) And you had earphones on, and you heard it over that?

(MICHELLE) Headphones. Yeah.

(Q) And you heard it over that?

(MICHELLE) Yeah. 'Cause there was just gobs and gobs of helicopters. I think I looked out the window first and then...

(ROSALIE) They looked like forest rangers. There were green, you know, it's according to how far off they were...

(MICHELLE) And some of them by the trees. One of them that wasn't sitting on the ground or nothing, but you could see its color. It was pretty close to the ground.

(ROSALIE) [Now outside pointing in the direction of the glow] There was this big glow just over the top of the tree line... I'd say it is about six miles across there. Right over my neighbor's house, between these two, and at the edge of the tree line is where it was.

[Crosby is located just south of the edge of the East Texas Piney Woods on the coastal prairie. The incident occurred in the woods to the north]. Rosalie described the locale as follows:

It wasn't all of the way over to the Cleveland highway. It was back toward my house more, between, y'know, between the highway going to Huffman right over in that area.

(Q) Can you describe the shape of the helicopters? Was there anything unique about them? Did you see any markings, or things like that?

(ROSALIE) No.

(Q) No markings?

(ROSALIE) Green, dark green, the one I...

(MICHELLE) They looked green.

(Q) Were they small or large? Or were there different kinds?

(MICHELLE) They seemed to be...

(ROSALIE) They were just helicopters.

(Q) Just helicopters?

(ROSALIE) Yeah. If you was to ask me what kind of car this red car is out there, I'd tell you it's a red car. I mean, I don't know models. And helicopters, you know, it looked like a helicopter. It was a helicopter.

Rosalie was concerned about how the helicopters could fly so low and not run into anything, so Marvin explained his theory about why they were flying that way.

(MARVIN) A helicopter can drop down just like that.

(ROSALIE) But there's power lines...

(MARVIN) There's trees between here and where all of this happened to that woman [Vickie]. There's a tree line out there. They wasn't fixing to come down to land here. They were evading radar is what they were doing.

(Q) Evading radar?

(MARVIN) That's what I figured. Now that we know what the hell went on, that's what I figured.

(Q) You mean they were staying below it so that they couldn't be seen?

(MARVIN) Right, that's what I figure. I just figured they were going out there to do something. You know, maybe some kind of military maneuver. But now that I see what's happened, it's pretty obvious.

Our cross-questioning continued, searching for more clues about what Rosalie and Michelle had seen.

(Q) Did you notice any lights on them?

(MICHELLE) They were shining all over the ground.

(Q) Searchlights?

(MICHELLE) Yeah, that's what caught my eye. 'Cause I heard 'em, and I looked out the window. And at first, we thought they were looking for somebody or a wreck or something, 'cause they had lights.

(Q) Oh, they had searchlights on the ground. Were they looking down?

(MICHELLE) They weren't flashing down on us, but just gobs of them...

(ROSALIE) Remember, Michelle, what we did? We waved...

(Q) You waved at them?

(ROSALIE) We sure did wave at 'em. And we thought they'd either wave or do something, but they didn't.

(MICHELLE) There was just gobs of 'em. That's excitement for Crosby, to have lots of helicopters.

Rosalie explained how some of the neighbors were gone for the holidays. She pointed out the significant distance between houses. Each lot is at least two acres, so when they are home and inside, they seldom see their neighbors anyway. Then Rosalie and Michelle began talking about the event all over again, noting the sound again and pointing out where all of the helicopters were flying.

(MICHELLE) I think I just heard the noise of it.

(Q) You didn't notice any influence on your radio? Nothing bothered your radio? No static?

(MICHELLE) I don't remember.

(ROSALIE) The music she listens to, she probably thought it was part of it. That's why she had headphones on, 'cause I can't take it.

(MICHELLE) I mean I don't ever come out of that room... I mean, I wouldn't just come out if I didn't hear it...

(Q) There was no particular direction or anything about them?

(ROSALIE) I thought they were coming in from all directions, mostly from Houston, and mostly, y'know like right over here and right over there [pointing northwest]. There wasn't none of them coming this way—we didn't see them coming from like Dayton [to the east]. They were all coming like from Houston, around this way. And some of 'em were, y'know, over on this side.

(Q) That would be more to the north of Houston then?

(ROSALIE) But most of 'em were right over here. Right on over in here [pointing to the southeast of the house]. That's where they looked just like they were coming from Houston, y'know, right across Lucy Bell's is where I first saw 'em. And like I said, then we got looking, and they were everywhere. And I'm sure the ones that we saw over here toward Alice's had come from that direction too, I guess. They were everywhere!

There was a discussion about how quiet the area normally is. Since it is an open rural community, there are no sounds of the city. Michelle used the sound of a train passing in the distance as we talked, as a comparison of how easy it is to hear sounds outside their house.

(MICHELLE) See, right now, you can hear something.

(Q) I can hear a train.

(MICHELLE) It's a train! So, I could hear all of them.

(MARVIN) And they were coming right over the house and everything, so they were pretty loud.

(Q) You said you checked the calendar, and you verified right on the calendar what day it was?

(MARVIN) Yes, we did.

The size of the helicopters was also discussed. There was no doubt, these witnesses were all impressed by the large size of the helicopters and the noise they made.

(MICHELLE) But we did see the helicopters... They weren't the kind that just had the little round dome things.

(Q) Oh, they weren't?

(MICHELLE) They were big. I don't remember, but it seems like they had doors. But I do know they weren't the little round dome kind.

(Q) They weren't the bubble kind?
(MICHELLE) No.
(Q) They were long?
(MICHELLE) Long. Yeah.

Michelle was very impressed by how close one of the large helicopters was to the house. She said one was hovering near a little tree in their yard, only 150 feet or so away.

(MICHELLE) I saw one that was over by that tree out there. That's the one I remember the most.

(Q) That close! I can imagine you would remember. Did you hear them anymore that night after that?

(ROSALIE) No.

(Q) You never heard them later?

(ROSALIE) No. We'd go out every once in a while, looking.

(Q) Did you really?

(ROSALIE) We wanted some more excitement, but we never heard any more after that. In fact, or since.

(MICHELLE) You know there's planes that come...

(ROSALIE) These were no planes.

(MICHELLE) They were helicopters.

(ROSALIE) It reminded me of that movie...

(MARVIN) Apocalypse Now.

(ROSALIE) The way those helicopters were coming in low like, y'know. That's exactly what it reminded me of.

(Q) How long do you think was the total time that you saw the helicopters? A minute? Five minutes? Ten minutes? Twenty minutes?

(ROSALIE) I'd say, maybe fifteen minutes.

(MICHELLE) And there were some over in here-just gobs of 'em.

(ROSALIE) And I remembered the lights on the helicopters. Like they were looking for something.

(MICHELLE) You know you always think when you see searchlights—oh, they're looking for some man running through the field. Well, that is what I think.

(ROSALIE) Yeah, sure, and they were even right over this field right in here [pointing southwest]. Some of them were kind of high, but a lot of them were real low.

(MICHELLE) They were all going this direction [pointing north, northeast].

They spent a lot of time discussing the helicopter by the tree, and we verified it was 140-150 feet from the house. Then they did a role-play to show us how they moved around outside the house to observe the helicopters in all directions.

(ROSALIE) And then there was one out there over that pond...

(MICHELLE) We started out there behind the carport [south side of the house]. And then we ended up right over there [pointing to the driveway in front of the house]. There were just gobs of them.

CHAPTER 31
Unconfirmed But Interesting Reports Of Helicopters

A lot of people have come forward with reports of helicopter activity that may be related to this incident. Some said they "believed" they saw helicopters that night but could not be sure. Others were sure it was the same night, but they did not want their names used in conjunction with their reports. Situations like this are always frustrating to the investigator and to the victims; however, the wishes of the witnesses must be respected.

One unconfirmed, but interesting report came from a co-worker in the aerospace business. An employee of AVSCOM, the Army facility in St. Louis, Missouri, related the following information about the failed raid into Iran to rescue the hostages. He said they used H-53 helicopters on the raid rather than CH-47s because the Chinooks didn't have the Omega system installed. The CH-47s were the Army's first choice for the mission and were flown on parallel missions in the United States during the training period and during the mission in Iran.

They were flying between Fort Carson, Colorado, and Fort Campbell in Kentucky. They flew at night arid at very low levels during the month of December 1980. He cited "possible" other long-distance flights in Texas but was afraid to say much more. Tom Adams, an excellent UFO investigator, writing his own publication known as *Stigmata*, issue No. 19, told of related information he had obtained from a military person. The following is an exact quote from Tom's report:

Project Stigma has a source in the military community whom we shall call "Victor." While being normally stationed in a particular area, Victor periodically visited yet another military base in the U.S. during 1989. While at this second base, he was contacted by another military man ("Tony") who had sought to meet Victor, knowing of the latter's interest in UFOs. He wanted to tell Victor of an experience he had had while stationed at Fort Hood, Texas, in December of 1980. While Tony related his story, Victor closely observed his facial expressions and noted every nuance in his tone and manner of speaking. Victor felt that Tony was being sincere.

Tony was a helicopter pilot at Fort Hood. Between Christmas and the end of 1980, there was a "special alert." The helicopter pilots (not known how many participated, other than Tony) did not know where they would be going—that once they reached the general area, they would be vectored in. They were told they would see an "unusual aircraft" there—that their mission was to "cover" this craft and try to force it to land by keeping it at a low altitude. Once the craft landed, they were to "mark" the spot, radio in, and wait for "other people" to move in. The choppers would then be dismissed to return to their base. Tony told Victor that the whole operation happened so suddenly that he was "half-drunk" when he took off from Fort Hood, but that what he saw later sobered him up. They reached the area in question, and Tony saw a craft he described as "the biggest damndest diamond he ever saw in his life." As the choppers flew over the top of the object, trying to "hold it down," it was "throwing off sparks like a 4th of July sparkler." Sometimes the sparks would hit the ground; sometimes, they would appear to burn out before hitting the ground. The craft began to move off, and the choppers followed it for 7 to 10 miles. As they were following it, it stopped throwing off sparks and started to "glow." The craft then became stationary again. At that moment, the choppers received word that they were to abort the mission and "return to their various bases" (i.e., they had come from more than one base). Later, the Fort Hood personnel who had been involved in the mission were re-assigned. Tony ended up in Germany.

After their arrival back at Fort Hood, everyone was debriefed. They were told that what they had encountered was an "experimental aircraft" that had gone astray outside of its flight pattern and had started to "experience problems." It was crucial to ensure that no one [civilians] got too close to it. Not all of the choppers were CH-47 "Chinooks;" some were UH-1, "Hueys," which is what Tony was piloting. The Chinooks had "contact teams" in them. If the craft had crashed or set down, these "contact teams" were to go down and secure the area. From what Tony had understood, there had been four contact teams.

Tony couldn't verify the total number of choppers [23?] since they had come from different bases. Tony had no idea what the "diamond object" really was. They were told they would be vectored into the location, but then they would have to "go visual" because the object they were looking for would be flying below the radar. Apparently, thought Tony, whatever malfunction had occurred on board the strange craft, it was corrected. Tony said he had kept his mouth shut about this incident, and nothing untoward had happened to him since then. He felt that Victor would be interested in the story, but Tony advised he should keep his mouth shut about it, as well.

CHAPTER 32
The Legal Battle Begins

In the beginning, Betty and Vickie had no intention of filing a legal claim against the United States Government. They believed, erroneously, that if they contacted government representatives, that information about the device that harmed them would be made available to their doctors so the proper treatment could be applied. After receiving no useful response to more than one hundred telephone calls and as many letters, it became obvious that help was not forthcoming.

Betty received a letter from the Center for UFO Studies dated May 31, 1981, suggesting it was time to submit a formal complaint about the experience. Fred Whiting, at the time a Congressional aide as well as a UFO researcher, had provided a set of guidelines to follow in formalizing a contact with Rep. Charles Wilson, Sen. Lloyd Bentsen, and Sen. John G. Tower. Thankful for Fred's help, we followed his suggestions to the letter. And his help did not end there. He tirelessly pursued disclosure of information about the incident through the Freedom of Information Act (FOIA). Representative samples of that material appear later in the book.

The first to respond was Rep. Charles Wilson. In a letter dated July 20, 1981, Representative Wilson said: "The military officials I contacted were not aware of any military operation that was being conducted in that area on December 29, 1980." His only suggestion was to contact Mr. Frank Stranges in Van Nuys, California.

A letter dated July 28, 1981, was received from Sen. Lloyd Bentsen. He told them to contact the Judge Advocate Claims Officer at Bergstrom Air Force Base in Austin, Texas, where they could file an official report and submit a claim. He said he had been told that those officials were aware of the incident and would be "most willing to assist in any way possible."

A meeting at Bergstrom was set for August 17, 1981. Betty, Vickie, and Colby went to that meeting with enthusiasm, believing that at last, they would receive information that would lead to relief from their maladies. Unfortunately, that was a bad assumption.

The trio was taken to a meeting room with a large map of the United States on the wall. Interestingly, a marker had been placed on the map noting the location of Huffman, Texas. The following representatives of the Air Force were present: Capt. William J. Camp, Acting Staff Judge Advocate, Capt. Terry Davis, Claims Officer, and Pat Wolfe, Assistant Claims Officer. Betty and Vickie gave the Air Force representatives copies of photographs of their injuries, names of their doctors, and information about the incident. A tape recording of the conversation was made and retained by the Air Force. Then they were "dismissed" by the Air Force people. On the way out, they were offered the forms for filing a claim; with the admonition, they should have a civilian attorney assist with the filing, "if they could find one that would do it." That arrogant response gave us all renewed vigor to seek justice in this case.

The next official communication was a letter from Sen. John Tower dated September 4, 1981. He said he had been in contact with "appropriate authorities" and that Betty and Vickie needed to file a claim with the Bergstrom AFB Staff Judge Advocate.

In the meantime, we had already been busy gathering the information required for the claim. Attorney William C. Shead, a MUFON consultant, offered to help; so, Betty, Vickie, and I met with him on August 23, 1981. He said not to be in a hurry because all of the backup documentation needed to be assembled before filing. He accepted the task of assembling a folio containing all of the facts to accompany the filing.

Soon afterward, a New York attorney, Peter A. Gersten, called me at work offering to handle the claim. Because of his excellent track record in gaining the release of previously classified UFO information through the FOIA, we all agreed he would be an asset to the team.

On April 8, 1982, Bill Shead transmitted medical records, letters, statements, and miscellaneous reports to Peter Gersten to be used in support of the claims to be filed under the Federal Tort Claims Act. Gersten worked on the folio for several months, culminating in his submittal of three claims to the Base Judge Advocate at Bergstrom, using the Department of Justice Standard Form 95 (Rev. 6-78). Betty's personal injury claim was for $10 million. The claims for Vickie and Colby were for $5 million each.

Although the claim had been filed at Bergstrom AFB as required, the reply on May 2, 1983, came from R. R. Semeta, Chief, Claims and Tort Litigation Staff, Office of the Judge Advocate General in Washington, D.C. His reply stated:

"Your clients' claims for personal injury allegedly caused by an overflight of an unidentified flying object and unidentified helicopters on 29 Dec. 80, have been considered under the provisions of the Military Claims Act, 10 U.S.C. 2733, and are denied."

"The reason for this decision is that the attendant facts fail to establish that the unidentified flying object or helicopters were owned or operated by the United States government or any agency or instrumentality thereof."

Colonel Semeta's response did leave the door open for an appeal to a higher authority within sixty days. Gersien followed through with submittal of the appeal on July 20, 1983, based on the following points:

(1) The object in question involved an experimental device which through guidance and/or propulsion trouble found itself far off range and crippled. A military rapid deployment team (the helicopters) was mobilized on an emergency basis to escort the troubled vehicle or to secure the area in case the vehicle was forced to land. Any such operation would be of high national security nature and not be subject to a routine disclosure.

(2) The object in question was a foreign aggressor, either terrestrial or extraterrestrial, similar to the object that was • observed at RAF Woodbridge, England on the nights of 27- 29 Dec. 1980.... Once again, any such encounter would be of a national security nature and not be subject to ordinary discovery.

Furthermore, it appears that my clients' observations of the existence of an unusual airborne object are corroborated by the enclosed Air Force document and other civilian reports of similar objects seen at about the same time.... Based upon the presence of the UFO and military-type helicopters and our inability to determine their nature and origin due to national security restraints, the burden of proof is now shifted to the government to prove that it is not responsible for the resulting injuries to my clients.

Please be advised that my clients have authorized me to initiate a lawsuit in the appropriate U.S. District Court and pursue any and all discovery procedures if there is no reasonable compromise and settlement which would honor the needs of all parties to this unfortunate situation.

The response to the appeal came on September 2, 1983, from Charles M. Stewart, Colonel, USAF, Director of Civil Law, Office of The Judge Advocate General in Washington, D.C. Although the appeal was denied as expected, it left the door open for further litigation. The reply stated the following:

"The appeals of your clients' claims for personal injuries allegedly caused by an overflight of an unidentified flying object and unidentified helicopters on 29 December 1980 have been considered under 10 U.S.C. 2733 and are denied."

"The reason for this direction is that the facts as alleged by the claimants fail to establish that their injuries were caused in any way by the United States Government or any of its agencies or instrumentalities. You should not consider the acceptance or denial of this claim as an admission of the truth of any facts alleged by your clients. Our investigation has revealed no evidence of involvement by any military personnel, equipment or aircraft in this alleged incident. The arguments you presented to establish liability of the government are not supported by any case or statutory law."

"This is the final administrative action that can be taken on your clients' claims. This denial also satisfies the administrative filing requirements of the Federal Torts Claims Act. Based on this denial, your clients have the right to file suit against the government in an appropriate United States District Court not later than six months from the date of the mailing of this letter of denial."

With the permission to file suit against the government in hand, the next step was to proceed with the filing.

CHAPTER 33
Federal court action begins

Peter Gersten took the first action in filing suit by preparing an "Amended Complaint" for "Civil Action, File Number H-84-348," claiming that the injuries to the victims were caused solely by agencies and employees of the U. S. Government without any negligence on the part of the injured parties. The Plaintiffs were identified as Betty Cash, Vickie Landrum and Colby Landrum and the Defendant was the United States Government. The documents were filed in the United States District Court for the Southern District of Texas, in Houston, in January 1984 by Bill Shead.

The complaint charged the U. S. Government with negligence and defined the extent of the negligence in fifteen numbered paragraphs. Extracts from the complaint follow:

During all times hereinafter mentioned, defendant owned and operated military CH-47 double rotary type helicopters and an experimental aerial device of a hazardous nature. At all times hereinbefore mentioned defendant did not use proper care and skill in failing to warn or protect plaintiffs from said experimental aerial device which was clearly hazardous in nature.

At all times hereinbefore mentioned, defendant negligently, carelessly, and recklessly allowed said experimental aerial device to fly over a publicly used road and come in contact with plaintiffs.

Solely by reason of defendant's carelessness and negligence as aforesaid, plaintiff experienced the following symptoms and injuries:

The U. S. Government was represented by United States Attorney Daniel K. Hedges and his Assistant Frank A. Conforti. In March Conforti filed a "Motion for More Definite Statement" with the Court, invoking Federal

Rule of Civil Procedure Rule 12 (e). He claimed that the "experimental aerial device" and "unconventional aerial object" statements made by Gersten were vague and ambiguous.

Conforti said the United States is unable to properly respond because: "the United States, as a part of its defense capability, uses an extremely diverse variety of aircraft, any number of which might be considered "unconventional." Further, the United States, both as a part of its defense development programs and as a part of such endeavors as the space program, designs and tests aircraft which might be considered "experimental."

The court responded with an "Order for More Definite Statement," giving Gersten only ten days to respond. The order asked for a more detailed description of the "unconventional aerial device," including shape, size, markings, sounds, smells, visual aspects, and other sensory observations.

Gersten responded immediately with a "More Definitive Statement." In the response, he established the use of "UFO" as a common term to describe the unconventional aerial device. Further, he reiterated the Plaintiffs' description of the object, along with its size, shape, and altitude. He described the flames, the glow, the heat, and the plaintiff's distance from the object.

On April 5, 1984, Conforti finally filed an answer to the charges, listing four lines of defense.

The first defense was to deny the charges as listed in all fifteen paragraphs of the complaint. In seven of the paragraphs, Conforti claimed having insufficient knowledge or information to form a belief, and this became a part of their denial. He concluded the first defense with the following statement:

"Finally, the United States denies that Plaintiffs are entitled to any of the relief requested in this action. All allegations in Plaintiffs' complaint not specifically admitted herein are denied."

The second defense stated: "Plaintiffs have failed to state a claim upon which relief may be granted."

The third defense tried to shift the blame for the injuries by stating:

"Plaintiffs, by reason of their actions and/or nonaction, assumed the risk of the injuries alleged to have been sustained."

Conforti is saying that it is their own fault they were hurt because they didn't get away. I believe by this statement he is also saying that even if the government did cause the harm, it is the plaintiffs' fault because they were there at the time. This could set the stage for later denial, just in case the government's involvement was actually proven.

The fourth defense built a case for negligence on the part of the injured parties:

"Plaintiffs, by reason of their actions and/or nonaction, were contributorily negligent, and by reason of such contributory negligence, "the injuries alleged were sustained." Here, Conforti is saying that any person caught up in a government operation is negligent just by being there. This attitude of non-responsibility on behalf of any government employee has frightening ramifications and should be exposed and refuted whenever it happens.

Conforti summed up his answer by requesting the Court "to deny Plaintiffs all relief requested."

CHAPTER 34
The Process Of Discovery

Gersten Asks The Questions

Using Rule 33 of the Federal Rules of Civil Procedure, Gersten filed "Interrogatories Number H-84-348," requiring the Defendant to answer specific questions related to the case, together with a statement identifying the source of such information.

Gersten began the Interrogatories with "Definitions and Instructions." In it, his definition of "documents" was the most comprehensive definition I had ever seen. Because of its potential use by other researchers, it is repeated in its entirety herein: As used herein, "document" includes raw data, research data, interview reports, books, records, correspondence, telegrams, interview notes, tabulations, compilations, charts, surveys, appraisals, worksheets, and other reports, letters, correspondence, notes, pamphlets, leaflets, diaries, telegrams, desk calendars, appointment logs, memoranda or oral conferences, memoranda of conversations, memoranda of meetings, memoranda of telephone calls. Minutes and all transcriptions or reproductions by any means thereof, together with all drafts of any written document, and all other documentary material of any nature whether written, printed, typed, recorded, or other graphic matter, however produced or reproduced in defendant's possession, custody or control, from whatever source obtained and whether or not prepared by defendants.

"Document" also includes all records showing the identities of organizations, consultants, scientific and technical personnel who conducted and collected or assembled data or participated in any manner in the preparation of studies, reports, surveys, appraisals, or evaluations.

"Document" shall refer to all originals and all non-identical copies, except that copies which differ by reason of notations made thereon are not considered identical copies.

Gersten gave clear instructions about how to identify persons, entities, organizations, and locations. He expanded on the definition of documents by describing what to include when identifying a document. He even gave a detailed definition of what he meant by the words "describe" and "explain." By doing this, Gersten eliminated any further claim by Conforti that the Government did not understand what was being requested.

Now that the definitions and instructions were understood, Gersten listed nineteen interrogatories, starting with: (1) What is a "CH-47" helicopter; followed by questions about where CH-47 were built, operated, and maintained; performance parameters; definitions of crews along with listings of all personnel qualified to operate CH-47s at that time and their medical records; and copies of flight plans and maintenance records.

He requested information, knowledge, and documents about the incident from thirteen different U. S. government organizations and asked if any government agencies had investigated the incident. Included was a request for the government to identify "Project Snowbird" and "Project Moondust" because these names had been suggested by other researchers as having significance in the helicopter operations.

Conforti Strikes Back

On January 17, 1985, Conforti filed a "Defendant's Motion to Dismiss And/or For Summary Judgment," pursuant to Rules 12(b) (1) and (6) of the Federal Rules of Civil Procedure. What he was asking for was an Order from the Court dismissing the suit on the grounds that the Court lacked jurisdiction over such action and the Plaintiffs failed to state a claim upon which relief could be granted. He also asked, under Rule 56 of the Federal Rules of Civil Procedure, for an Order granting Summary Judgement in favor of the government. He devoted a lot of time to the discussion of the "UFO", but avoided discussion of the more than twenty helicopters, apparently done in a move to distance the government aircraft operations from the "UFO" activities.

In support of his position, Conforti submitted a "Memorandum in Support of Defendant's Motion to Dismiss and/or For Summary Judgement." The

Supporting Memorandum included the following statement establishing as a part of the request for dismissal, expert testimony:

Filed herewith are the sworn affidavits of Robert W. Sommer, NASA; Colonel William E. Krebs, USAF; Vice Admiral Robert F. Schoultz, USN; and Richard L. Ballard, Office of the Deputy Chief of Staff for Research, Development and Acquisition, USA. The affidavits establish that the "UFO" allegedly seen by plaintiffs, and which it is alleged was the proximate cause of their asserted injuries, is not, and was not, owned, operated, or in the aircraft inventories of the United States of America nor was such an object under the control of the United States of America or its employees.

On the basis of those affidavits, the United States moves this Court for an Order dismissing the Complaint of plaintiffs with prejudice, or, in the alternative, finding that there exists no genuine issue of material fact for summary judgment in favor of the defendants.

Conforti then presented an argument in support of his contention that the government was blameless. He said: "Under the Federal Tort Claims Act, the question of liability is determined by reference to the law of the state in which the alleged tortious conduct of the defendant, in this case— negligence-occurred... Accordingly, the determination of whether the United States was negligent herein must turn upon the prerequisites for a negligence action in Texas. Under Texas law, a plaintiff must prove the existence of a legal duty owed to him by the defendant in order to establish tort liability. In the absence of such a legal duty or of injury from its breach, there can be no actionable negligence and hence no legal liability. The existence of a defendant's duty-is a matter of law, distinct from factual matters of breach and consequence."

Further, Conforti stated: "The position of the defendant, United States of America, is that plaintiffs have not shown, and cannot show, the existence of a legal duty owed to them by the defendant. Hence, plaintiffs have failed to state a cause of action under the Federal Tort Claims Act for which recovery may be granted."

On the question of operator or ownership of the "UFO," Conforti stated: "...the United States neither owned, operated, nor controlled the alleged "UFO". As such, it is axiomatic that no legal duty may result which is attributable to the United States. Nor may actions or omissions, if any, of employees of the United States result in liability."

As if that was not enough, Conforti denies any responsibility even if the government-owned the "UFO." He says the government can do whatever it wants and cannot be held liable, as follows:

"Assuming, arguendo. that the United States owned, operated, or otherwise controlled, the "UFO," plaintiffs assert that the government negligently permitted the "experimental aerial device" to fly over a public road and failed to warn the plaintiffs that the "experimental aerial device" was clearly hazardous in nature. With respect to the alleged hazardous nature of the object, it is settled law that *the United States may not be held strictly liable for undertaking an ultrahazardous activity.*

Conforti also addressed the issue of the government's "failure to warn" by showing that the government's actions were proper within the "discretionary function of the government." He draws heavily on the fact that "UFO" is applicable to an object which is not known or identifiable, and as such, the government could not have known whether any danger existed or from whence such danger could spring. Then he addresses the reasonableness of the government's actions, stating that Texas law requires that the emergency nature of the situation must be considered. He said: "Faced with the situation of an unknown object, a governmental determination *not to issue a warning, and potentially cause a panic with the known dangers arising from a panic,* simply would not constitute negligence in any event."

Even if a UFO was involved in the incident, Conforti says the actions by the government were still proper. His reasoning leads one to believe that the government used a standard operating procedure in dealing with the UFO. Conforti said: "Assuming, arguendo. that the plaintiffs were correct in their assertion that the object they may have seen was a "UFO," what could be more of a discretionary function that a decision by the United States and its armed services *whether and how to react?* It must be recalled that plaintiffs, themselves, concluded that military aircraft were escorting and surrounding the "UFO." If, as must be done in a motion such as this, the pleadings of the plaintiffs are accepted as true, then plaintiffs themselves have made. the government's case for application of the discretionary function exception the weighing of governmental interests and *deciding in favor of the less antagonistic approach,* clearly constitutes the type of-discretion reflected in the history."

The Government's Expert Witnesses

Gersten had been quite clear in his "Interrogatories" that he expected statements from thirteen agencies, including the Rapid Deployment Force, Army Intelligence, National Military Command Center, and more. These were seen as agencies with an interest in and knowledge of the subject operations. Expert statements from these agencies and the Nuclear Emergency Search Team could bring relief to the plaintiffs.

Instead of responding to the Interrogatories as normally required by law, the government chose to select some "other" carefully chosen "experts" in its response. While they all appear to be highly qualified individuals in their respective areas, the government's experts didn't appear to have knowledge about what went on in this incident and they didn't answer the Interrogatories. In other words, they gave truthful but safe answers.

Gersten had not requested information about NASA operations because we already knew that NASA doesn't fly large groups of helicopters. Conforti's first expert witness was Robert W. Sommer, NASA Deputy Director of the Aircraft Management Office. He testified that NASA had one CH-47 helicopter located at Ames Research Center in California. It was not flying on December 29, 1980. Further, he said: "I declare that no 'object' as described by plaintiffs was, at any time, owned or operated, or was in the inventory or under the control of NASA." Personally, I agree with Mr. Sommer's statement, because I have seen no evidence that NASA is flying UFOs. However, if Conforti really wanted to find out what type of new research spacecraft NASA was developing, Sommer would not have been the correct NASA expert to choose.

While Sommer could give a true and proper assessment of NASA's aircraft operations, he was not the expert in advanced spacecraft development. That function does not and never has come under the jurisdiction of the NASA Aircraft Management Office. It is the responsibility of the NASA Advanced Projects Office. Conforti "used" NASA's name and reputation as the Nation's premier space organization to give a plausible-sounding answer, without danger of exposing anything related to the case.

The second expert witness was Air Force Col. William E. Krebs, Chief of Tactical Aeronautical Systems Division. He declared that "the CH-47 was not in the inventory of the U.S. Air Force...." Again, Conforti had used another "aircraft" expert and did not expect him to know about the UFO. Clearly, Krebs had nothing to do with the Air Force Space Command. He was an "airplane" expert. In his testimony, he compared the description of the "UFO" with known Air Force aircraft and did not find a match. He said:

"I have compared the description of the object with my knowledge of the inventory of all United States Air Force craft capable of flight. No such craft was owned, operated, or in the inventory of the United States Air Force on December 29, 1980." His words were carefully chosen and no doubt true as he stated them. However, no testimony was given by any of the Air Force agencies requested in Gersten's Interrogatories - agencies with knowledge of covert operations.

The third witness was Navy Vice Admiral Robert F. Schoultz, Deputy Chief of Naval Operations (Air Warfare). Potentially, this expert could have addressed important material such as operations of amphibious assault ships, helicopter aircraft carrier operations and the operations of double rotor helicopters such as the CH-46 and CH-47; but he didn't touch on any of this. Instead, he only compared the description of the "UFO" with naval aircraft. He said: "....no aircraft matching the description given in Exhibit A was owned or operated by the United States Navy on December 29, 1980, and no such aircraft is currently owned or operated by the United States Navy." No one had suggested that the "UFO" was a naval aircraft. The Admiral answered a non-question and carefully avoided the important issues.

Conforti rounded out his file of expert testimony with a statement from Richard L. Ballard, Acting Chief of the U. S. Army Aviation Systems Division. Again, this "aircraft" expert addressed whether or not the "UFO" was an Army aircraft and completely avoided discussion of the helicopters. He said: "I have compared the description of the object in Exhibit A with my knowledge of the inventory of all Army craft capable of flight. No such craft was owned, operated, or in the inventory of the United States Army on or about December 29, 1980." This expert could have answered the long list of questions about CH-47s as requested by the Interrogatories but failed to do so.

While statements by other experts could have been used, Conforti did a masterful job of selecting "safe" testimony. This testimony would impress the Court because the statements were from the three major armed forces and the space agency. It didn't matter that the testimony either failed to answer the Interrogatories or that the particular expert was not from the correct part of the agency to have full knowledge of events. He had avoided the real issues and got away with it.

The Docket Call

After Bill Shead had made a lot of visits to the Federal Court Building in Houston and Peter Gersten had prepared the paperwork for Bill to file with the Court, we were pleasantly surprised when on January 31, 1985, the case was set for Docket Call on September 3, 1985, at 11:00 a.m. in the court of United States District Judge Ross M. Sterling.

Betty and Vickie were happy for the first time in several years. They believed they were finally going to get some fair treatment. To them, appearing before a Federal Judge represented the highest level of truth and honesty they could hope for under the American judicial system. It gave them something to live for the next eight months.

CHAPTER 35
The Interrogatories Are Poorly Answered

In a surprising move, Conforti sent a document entitled "Answer of Defendant to Plaintiffs' Interrogatories," to Gersten on March 18, 1985. The answers, however, turned out to be almost useless. Conforti had done a clever job of truthfully answering the easy questions such as what is a CH-47 helicopter? The more important questions were either only partially answered, or not answered, or untruthful answers given. The quality of the answers was far below what the U. S. Government was capable of giving. For example:

QUESTION NO. 3: State the number of CH-47 helicopters in operation in December 1980.

ANSWER NO. 3: There were 450 CH-47 helicopters fielded in December 1980. An unknown, but certainly smaller, number were "in operation."

A simple analysis of this answer shows it was evasive and not complete. Of the several thousand CH-47s built, Conforti accounted for only 450 of them. Since the government keeps extensive records on every aircraft in the inventory, there had to be accountability for all CH-47s, as well as how many were in operation. It is obvious that Conforti knew that if he said how many were "in operation," that we would have expected to see where and when. Hence, he did not answer the question.

QUESTION NO. 4: State the distribution and location of all CH-47 helicopters in operation in December 1980.

ANSWER NO. 4: CH-47 Aircraft were distributed at the following locations in December 1980: Twenty locations were listed.

This answer was incomplete in two ways. First, it did not list all of the locations. Among those locations, not listed was Ellington Field - in Houston, Texas. Major Haire's earlier testimony established that eight CH-47s were there. In defining the distribution, Conforti avoided listing how many helicopters were at each base. If he didn't understand what distribution means, he could have consulted Webster's Dictionary, which describes "distribution" as "frequency of occurrence." He also avoided stating the "location" of the helicopters on December 29, 1980. The answer to question no. 4 was nothing but useless information.

QUESTION NO. 5: State the performance capability of CH-47 helicopters including, but not limited to, range and fuel capability.

ANSWER NO. 5: Performance data for the "A" model CH-47 is shown below:

I won't repeat the data given because the only useful information was that the range of the CH-47 "A" model was 250 miles. And this was different range data than we had been given by other experts.

Interestingly, the main site for "A" models was Ellington Field and Conforti had not listed that site in the distribution of CH-47s in Answer No. 4. The sites he did list, such as Fort Hood, Fort Bragg and Fort Campbell, were flying newer models; but Conforti did not supply the performance capability for CH-47s at any of the main bases. Hence, the answer given to question no. 5 might fool the casual reader but would never be accepted by experts.

QUESTION NO. 6: State the number of personnel required to operate a CH-47 helicopter, and the duty and responsibility of each.

Conforti provided a precise answer to this question, stating that the minimum crew "under normal conditions" is a pilot, copilot, and flight engineer. He describes the function of each person. He did not address the size of the crew for anything other than "normal operations." Again, he avoided the subject of the interrogatories.

QUESTION NO. 7: Identify all personnel qualified to operate CH-47 helicopters prior to January 1981.

It appears we really struck a nerve with this question because Conforti wrote five pages of double talk on why he could not and would not answer this question. Had he answered truthfully, we would have had access to the pilots flying during the subject exercise. Conforti said it would be "oppressive, unduly burdensome and expensive to answer." There was no thought about how oppressive the issue was to Betty, Vickie, and Colby and how expensive it had become.

Conforti made a big deal out of what "qualified to operate a CH-47" meant and subsequently avoiding giving any useful answer. He did say, "The Army maintains no single record, computerized or otherwise, that contains the information possibly sought by the question." It is amazing the United States can have any semblance of organized operation of the armed services if there is no personnel database available to America's military leaders, telling them who their pilots are. This is beyond belief.

After stating that there may be 2,000 to 3,000 pilots who have been trained by the Army to fly CH-47s, Conforti admits there is a computerized list of pilots that could be screened. The several paragraphs that follow are aimed at making the issue appear too complex to answer; when in fact, the whole case is about the incident of the night of December 29, 1980. Nevertheless, Conforti got by without giving a useful answer.

QUESTION NO. 8: State the flight plans for all CH-47 helicopters on December 28-31, 1980.

ANSWER NO. 8: Flight plans are destroyed after 30 days.

Flight plans for 28-31 December 1980 are not available.

The answer may be true for all FAA flight plans, but the records of a major military operation would not be destroyed. They may not be available because they are classified or are marked "Exempt from FOIA," but certainly not destroyed.

QUESTION NO. 9: State whether the maintenance records of all CH-47 helicopters in operation and use in December 28-31, 1980 are available.

ANSWER NO. 9: Maintenance records for 28-31 December are not available. All maintenance records are destroyed after 6 months.

This answer is a lie. An aircraft log and maintenance file is kept on every aircraft in the inventory until it is finally scrapped. These files are used regularly in accident investigation cases.

QUESTION NO. 10: State whether the medical records for all personnel qualified to operate CH-47 helicopters prior to January 1981, are available.

ANSWER NO. 10: Unknown at this time.

Conforti wisely chose not to answer the question about CH-47 flight crew medical records. We believed that some of those crews may have been harmed during the event as a result of their closeness to the "UFO," in some way similar to the way Betty, Vickie and Colby were harmed. Armed with medical records, we may have had a direct line on any pilot who flew near the object.

QUESTION NO. 11: State whether any of the following agencies of the defendant have any information, knowledge, or documents concerning the incident referred to in plaintiffs' amended complaint:

a. Department of Energy's Nevada Operations
b. Air Force Inspection and Safety Center (AFISC)
c. The Army agency responsible for aviation safety
d. Aerospace Rescue and Recovery Service (ARRS)
e. Secretary of Defense
f. Joint Chiefs of Staff
g. National Military Command Center (NMCC)
h. Rapid Deployment Force (RDF)
i. Air Force Intelligence
j. Army Intelligence
k. Air Force Office of Legislative Liaison
l. Air Force Inspector General
m. Army Inspector General

ANSWER NO. 11: c. No, j. No, m. Yes

Whether truthfully or not, Conforti only answered the question for three of the agencies requested. This is a sin of omission in this case. It is especially troublesome since the reply to the next question depended on a complete answer to question no. 11.

QUESTION NO. 12: If the answer to "11" is yes, identify any and all documents, and state the nature and substance of any knowledge.

ANSWER NO. 12: The Department of the Army Inspector General has records relating to his inquiry into whether the Army, Army National Guard, or Army Reserve helicopters were involved in the incident alleged by plaintiffs.

In this answer, Conforti only partially answered the question with his statement about the Army Inspector General. Even then, he failed to provide "the nature and substance" of the knowledge. Apparently, Conforti considers his non-response to the question about a majority of the agencies listed in question no. 11, as meaning he did not need to cover them in question no. 12 either.

QUESTION NO. 13: State whether any agency of the defendant conducted an investigation into the incident....

ANSWER NO. 13: Yes.

This was a follow-on to question no. 11, intended to seek information about which agencies did investigations.

QUESTION NO. 14: If the answer to question "13" is yes, identify the agency.

ANSWER NO. 14: United States Army

QUESTION NO. 15: If the answer to question "13" is yes, state whether any documents, tape recordings, notes, photographs, scientific reports and other materials exist.

ANSWER NO. 15: Yes

We must assume that the "yes" answer means that the United States Army has all of the items listed in question no. 15. Unfortunately, Gersten's question was not worded properly to get copies of the available materials.

Conforti's answer to the Interrogatories stopped with question no. 15. He ignored the rest of the questions, but they are listed below for the record:

QUESTION **NO.** 16: Identify "Project Snowbird," and "Project Moondust."

QUESTION NO. 17: State whether the Nuclear Emergency Search Team (NEST) was involved in any operations in December 1980.

QUESTION NO. 18: If the answer to question "17" is yes, describe each operation including, but not limited to its location.

QUESTION NO. 19: State whether the Air Force Inspection and Safety Center (AFISC) keeps reports for incidents involving classified experimental aircraft.

We had reason to believe, as a result of information gained from other researchers, that questions no. 16-19 would provide important insight into what took place on the night of December 29, 1980. Apparently, Conforti understood that fact and avoided the risk of exposure by ignoring the questions. He obviously believed that the Court would never compare the list of questions with the listed answers, and he could get by with the omission. He was right. His answers to the Interrogatories provided no new information, and he successfully skirted the issues.

CHAPTER 36
The Docket Call

Preparation For The Docket Call

Conforti had submitted a listing of Interrogatories also, consisting mainly of questions about the injuries and requesting names of doctors and hospitals. Betty questioned Gersten about how she should react to the questions. Apparently, Gersten told Betty to go ahead and send the material if she had the energy to do it. So, Betty and Vickie assembled their medical records and lists of doctors and sent it off to Conforti. He never responded.

Just ahead of the September 3, Court Docket date, Gersten, through Shead, filed a "Reply to Defendant's Motion to Dismiss and/or For Summary Judgment." The document was short and to the point. It refuted Conforti's request, as follows:

I. The complaint filed by plaintiffs, when viewed in the light most favorable to the plaintiffs, states a claim against the United States upon which relief can be granted.

II. Plaintiff's claim is not barred under the discretionary function exception to the Federal Tort Claims Act.

III. There exists genuine issues of material fact.

Gersten's accompanying Memorandum of Points and Authorities clearly refuted Conforti's rationale for dismissal. In section I, Gersten stated that the United States was negligent in its operations and failed to warn the plaintiffs. Section II described what happened that night and how the plaintiffs were harmed. Further, it told how they sought relief through a claim against the United States and were denied relief. This resulted in the suit being filed. Section III defined the issues in the case, and Section IV provided the arguments in three sub-sections as follows:

The complaint filed by plaintiffs states a claim against the united states upon which relief can be granted.

A. There is only one aerial object referred to as a "UFO."

Defendant, in its motion to dismiss, states that "Plaintiffs imply though it is nowhere asserted that the United States owned and operated the 'UFO.'" Plaintiffs allege in their complaint that the defendant owned and operated an "experimental aerial device". It is clear from reading of the complaint in conjunction with plaintiffs "More Definitive Statement" that only one aerial object is involved. An object, because of its unusual characteristics, defies precise identification. The object is indeed aerial and unconventional and, from all appearances, experimental. The term "UFO" is used to avoid the possibility of mischaracterizing the object. The defendant misuses the term to create the impression that no triable issue of fact exists as to whether there existed a legal duty to plaintiffs by the defendant.

B. There existed a legal duty owed to the plaintiffs by the defendant.

Defendant contends that there existed no legal duty by defendant to plaintiffs and thus no claim can exist against the defendant. It is clear that if the defendant either owned, operated, or controlled the "UFO," there would exist that legal duty (the discretionary function is discussed in point II). Though the existence of a legal duty is a matter of law, the issue of whether the defendant owned, operated, or controlled the "UFO" is a question of fact.

Thus, assuming that the "UFO" was an experimental aerial device, one can infer that the defendant owned, operated, or controlled the "UFO" and the "UFO" would have the highest security classification.

The- defendant contends through affidavits that "the United States neither owned, operated nor controlled the alleged 'UFO.'" Defendant's affidavits are insufficient and should not be considered on the issue of ownership, operation, or control of the "UFO". *Nowhere in any of the affidavits do the deponents assert that they had the security clearance necessary to obtain this highly classified information. Each and every affidavit offers unsworn, self-serving opinions which are not supported by evidence of the nature and extent of the various searches for "UFO" information.*

Plaintiffs contend that these affidavits have failed to eliminate all triable issues of fact as to ownership and control. There were approximately two dozen military helicopters, including double rotary CH-47's, in the vicinity of the "UFO." Both the U.S. Army and the U.S. Marines have sufficient number of CH-47's to accommodate the plaintiff's observations. The presence of the helicopters is further evidence that contradicts the defendant's affidavits. Only in a trial with the right to confront and cross-examine witnesses can plaintiffs effectively explore and resolve these issues of fact. How can the defendant deny ownership of this "UFO" without being compelled to reveal the true owner of this clearly hazardous device?

The defendant would have us believe that the "UFO" was a foreign invader or possibly an extraterrestrial visitor, inferences that would bear more weight if substantiated by evidence. There is only one inference that can be drawn from the facts and circumstances of this case, the "UFO" was owned by the defendant. There are no other reasonable hypotheses.

Plaintiffs' claim is not barred under the discretionary function exception to the federal tort claims act.

Assuming that the "UFO" was owned, operated and/or controlled by the defendant, the only reasonable assumption in light of the defendant's lack of an alternate solution, then the negligence attributable to the defendant is in allowing this object to come over a public road. It is clear that this negligence is on the operational level and not the policy level and thus questionable as to whether it falls within the discretionary function exception. Furthermore, it is contended the defendant was negligent in failing to warn plaintiffs of this hazardous device, such failure not coming within the discretionary function exception.

The defendant created the danger by allowing this object to come over a public road and in contact with the plaintiffs. It is difficult to believe that the defendant is shielded from responsibility when a clearly hazardous device comes into contact with civilians over a public highway with the military presence and doing nothing.

There exists genuine issues of material fact.

Assuming, *arguendo* that the defendant did not own, operate, nor control the "UFO," a legal duty may still be attributable to the defendant. The Restatement (second) of Torts recognizes the duty to take affirmative action, including a warning. Section 322 of the Restatement provides that:

If the actor knows or has reason to know by his conduct, either tortuous or innocent, he has caused such bodily harm to another as to make him helpless, and in danger of further harm, the actor is under a duty to exercise reasonable care to prevent such further harm.

Plaintiff contends that there existed a limited legal duty owed to the plaintiffs by the defendant while the "UFO" was over the public road on which the plaintiffs were traveling. The presence of the helicopters implies knowledge on the part of the defendant of the existence of the "UFO", knowledge not only of the object but also of its dangerous propensities and its proximity to the plaintiffs.

Plaintiffs contend that this knowledge of the "UFO" threatening death or great bodily harm to another, which the defendant might avoid with a little inconvenience, creates a sufficient relationship recognized by every moral and social standard to impose a duty of action.

In this case, not only did the defendants vis-a-vis the helicopters take no action to avoid the danger to the plaintiffs, but the defendant also at no time attempted to warn the plaintiffs. The "UFO" was obviously a peril, not only threatening but actually causing great harm to the plaintiffs.

Assuming, *arguendo,* that the "UFO" was a true unknown, as implied in the defendant's motion, then the United States clearly has known about the obvious threat that "UFOs" pose for at least 35 years. It must have been this knowledge that warranted the presence of the military helicopters.

The plaintiffs pay taxes to the defendant for national defense. Obviously, this imposes some type of relationship which reaches the degree of legal duty when the defendant not only knows of the danger but has its military presence at that danger and then does nothing to warn the plaintiffs.

The day of the docket call

We arrived at the U. S. District Court building at 515 Rusk in Houston early in order to process through the security systems and get to the courtroom on time. Our party consisted of Betty Cash, Vickie Landrum, Vickie's daughter Jean Roper, Bill Shead, and me. Bill Shead was representing the plaintiffs in the case since Peter Gersten was unable to be in Houston for the Docket Call. The United States was represented by Frank Conforti and Judge Ross Sterling.

Judge Sterling immediately asked why Attorney Peter Gersten was not present, and Bill Shead said he was serving as the "local attorney for the plaintiffs" and was ready to proceed.

Sterling then asked: "How many days will it take to try the case?"

Conforti said: "I have no idea. Gersten hasn't responded. I have taken some discoveries. There is some mix-up between Gersten and his clients. He did not list the medical experts and presented only a partial list of the witnesses."

Shead said: "We still need some added discovery. The liability issue must be considered."

Sterling seemed impatient and said he wanted to rule on the whole thing. He said: "Gentlemen, I feel this case is almost over." He said he would rule on it soon and dismissed everyone.

We were all shocked. Not only did Judge Sterling not hear any of the evidence, but he also wasn't interested in seeing it and didn't set a date for court action. Later in the day, he appeared on a local television news spot and made a tongue-in-cheek statement about the case. It seemed to be a big joke to him.

Vickie states her feelings

Vickie had been keeping a log of her feelings, activities, and health state dating back to soon after the event occurred. It was one of the things I requested from her, and she did it faithfully. The following excerpts from her log, made right after Judge Sterling sent us away, serve to show the extent of her misery:

When Judge Ross Sterling said in the courthouse on September 3, 1985, he was "almost sure the case was over" before it was ever started, he never did review the evidence or hear us. He never looked at anything or heard nothing, for he would not let anyone say anything but Assistant Attorney Frank Conforti. My heart stood still like I could not breathe. I felt my life was over, and I could not move, for he was saying my case was over and I had no way to pick up my life or no one to turn to.

John Schuessler could not believe it either, for he was sitting beside me. John got up and began moving toward the door. I said to myself, may God have mercy on me and help me to do what I have to do. I got up slowly and walked toward the door. I looked up and saw my daughter Jean, and I think that in her face, I found faith to face what I knew I had to do. Slowly I walked up to Jean and said, come on, I need your help. I walked in front of John

Schuessler and Bill Shead, my lawyer, and he said: "Judge Sterling has made up his mind, so I don't know what else to do."

I said: come on, Jean, I need to hold on to you. I fought the tears back and prayed for strength. I rode the elevator down (from the ninth floor), and when the door opened, I stepped out and began walking toward the front. When I got outside is when the questions began coming from everyone at once. I tried to answer everyone with truth and dignity. I was asked how I felt now that it was about over. I told everyone that it was not over; the fight had just begun. If the Judge was a "just" Judge, he would give me a chance to be heard, for everyone in the whole United States, under the Constitution, should be heard.

I was wrong! For outside the Federal Court House in Houston, Texas, my civil rights were taken away from me that day. And until I have a chance to be heard in a "fair" court, I have no rights. I am praying for my rights back someday.

When I got back home and walked in, I knew everything was real, but it didn't seem so. This is the first time I ever lost faith in my God in my whole life. I thought He had turned against me, and everything didn't make sense. I had a feeling of loss I could not describe. I couldn't even talk to Colby. I felt so sorry for him. He would sit where he could touch me and beg me to tell him what was wrong. I had nowhere to go and no one to turn to.

Truly, justice had not been served that day in the court of Judge Ross Sterling. A lot of Americans lost their rights that day, and they didn't even know it was happening.

CHAPTER 37
Wrapping Up The Case

Vickie was right. We were not about to give up just because Judge Sterling failed to set a docket date. However, it was the next day, September 4, 1985, that we found out from Judge Ross Sterling's Clerk, Angie Siera, that they had not accepted our Reply to the Government's Motion to Dismiss and/or for Summary Judgement. Although it had been properly date stamped by the Court's Clerk on August 30, they were not taking it to the Judge because there was no evidence it had been provided to Frank Conforti.

Bill Shead returned to the Court on September 5, 1985, to present a Certificate-of-Service, attesting he had personally provided a copy of the response to Frank Conforti, the Attorney in charge for the Defendant. It was at that time he found that Conforti had filed a "Reply" on September 4 anyway, refuting Gersten's statements. It was a weak but effective document, considering the timing of the response, just one day after the Judge's little show.

Conforti said the plaintiffs were "attempting to confuse the basic issue of the Motion by setting up a "strawman" distinction between the "experimental aerial device". and the "UFO" He concluded: "that the United States did not own, operate, control, or otherwise contain in the inventories of its armed forces, the object characterized, by plaintiffs, as a UFO. Hence, under Texas law, no duty existed for the United States."

Then Conforti said: "Plaintiff tries to brush aside the four affidavits submitted by the defendant, asserting that the affiants did not have "the security clearances necessary to obtain this highly classified information." He admits that the four individuals were "responsible for aircraft inventories," but avoids the issue of new development projects where a device like the one in question might be found.

And finally, Conforti addressed the "discretionary function" issue, admitting that even if the Government was responsible for an experimental flight, the case could be dismissed under the discretionary function. He carefully ignored all of Gersten's arguments. He also addressed the "duty to warn" under the same "discretionary function" umbrella.

It was obvious by this time that the facts did not matter. Conforti was issuing the paperwork necessary for Judge Sterling to be able to dismiss the case without a hearing.

It was at this time that Bill Shead introduced Houston attorney Rhonda S. Ross to Peter Gersten as an interested lawyer to help with the case.

On February 17, 1986, Ms. Ross submitted to the Clerk of the U.S. District Court the "First Amended Motion to Continue Defendant's Motion to Dismiss and the Affidavit of Opposition to Plaintiff's Motion to Continue Defendant's Motion to Dismiss. Frank Conforti was notified of the filing at the same time.

Shead and Ross then filed a "First Request for Production of Documents of Plaintiff, Betty Cash et al, Addressed to Defendant." This move, pursuant to Rule 34 of the Federal Rules of Civil Procedure, asked the United States to produce, for inspection and copying, the documents that had been under discussion for many months. This action was a part of the incomplete discovery process that Shead had mentioned to Judge Ross Sterling during the September 3 Court Docket Call.

The action was no fishing expedition. Shead and Ross were going for the specific documentation that would expose the helicopter operations. They were specific in the meaning of "documents." They bounded the period of time covered by the request as: "....from January 1, 1980, to the date the documents are produced, or which relate to that period of time." They noted that some documents might be considered "privileged," and if so, they provided instructions on how to work with that categorization. And finally, they identified the facilities from which the documents were requested as: "....Fort Hood, Fort Sill, Fort Polk and Navy vessels capable of carrying and launching CH-47 helicopters in the Gulf of New Mexico on the dates noted." The documents requested were as follows:

1. Names and addresses of any and all temporary duty officers and/or

other officers assigned by and other means, at the above listed facilities for December 27, 28, 29, 30, and 31 of 1980.

2. All personal flight records for the officers listed in response No. 1 for the dates as noted above.

3. Copies of any documents that would reflect orders, plans or assignments for pilots of CH-47 helicopters to tow, ferry, and/or escort any large object through the air in December of 1980.

4. Serial numbers by type and model of all helicopters, of any type, assigned to the posts listed above or on temporary duty to the post listed above, or loaned to post listed above for the period of November 1, 1980 to March 1, 1981.

5. For each of the helicopters whose serial numbers were provided in response to the above request provide the material readiness reports.

6. Provide accountings from each of the posts listed above for fuel requested and/or supplied to the helicopters whose serial numbers were provided in request for production No. 5 above during the period from December 15, 1980, through December 31, 1980.

7. Provide lists of all serial numbers of all helicopters that were used to airlift or escort any objects from the posts listed above during the period from December 1, 1980, through December 31, 1980.

8. Provide the names and addresses of all enlisted men from the above listed facilities who flew any type of helicopter from December 27, 1980, through December 31, 1980.

9. Provide construction modification documents for the last ten (10) years for all underground facilities at Fort Hood, including the former Gray Army Airfield.

Shead and Ross also dealt with Conforti's request for dismissal in a very specific and professional manner, with the filing of: "Plaintiffs Response to Defendant's Opposition to Plaintiff's First Amended Motion to Continue Defendant's Motion to Dismiss."

This was a request for an order granting an extension of time to conduct Discovery, and it was well referenced to precedent-setting cases that supported the plaintiffs' rights to a favorable response. Excerpts from that filing illustrate the importance of continuing Discovery, as follows:

1. Plaintiffs Betty Cash, Vickie Landrum, and Colby Landrum have been seriously injured by radiation.

2. There is nothing, and there has been nothing natural in the area of FM Road 1485, 7 miles outside of New Caney, Texas, that would cause such severe and debilitating injuries.

3. One set of Interrogatories has been filed with little information discovered. Those questions were propounded by the Plaintiff's first attorney, and it is agreed that the questions in that first set were overbroad. New Interrogatories and a Request for Production have been drafted with special attention paid to the area of the United States in which the incident in question occurred. Special attention has also been paid to the type of documents to be produced. When the requested information is produced it will indicate that the Plaintiffs have stated a viable cause of action upon which relief may be granted and that the military operations in question did in fact occur under the direction of the United States of America on December 29, 1980.

4. Discovery will allow the Plaintiffs to prove that Plaintiffs' claims are not barred by the discretionary function exception of the Federal Tort Claims Act.

5. The evidence will show that the Plaintiffs have stated a claim upon which relief can be granted.

6. A dismissal under Rule 12(b)(6) of the Federal Rules of Civil Procedure is on the merits and is accorded a res judicata effect. For this reason, dismissal under this section is generally disfavored by the courts, and dismissal is proper only in extraordinary cases.

7. The burden of demonstrating that no claim has been stated is upon the Movant. In determining the motion, the court must presume all factual allegations of the complaint to be true, and all reasonable inferences are made in favor of the non-moving party.

8. Generally, the allegations of a complaint are to be liberally construed the (1983) Supreme Court stated:

When a federal court reviews the sufficiency of a complaint before the reception of any evidence either by affidavit or admissions, its task is necessarily a limited one. The issue is not whether a plaintiff will ultimately prevail but whether the claimant is entitled to offer evidence to support the claims. Indeed, it may appear on the face of the pleadings that a recovery is very remote and unlikely, but that is not the test. Moreover, it is well established that, in passing on a motion to dismiss, whether on the ground of lack of jurisdiction over the subject matter or for failure to state the cause of action, the allegations of the complaint should be construed favorably to the pleader. 416 U.S. at 236.

9. In making this determination, the likelihood that the plaintiff will prevail is immaterial.

10. Defendant's statement that Plaintiff is attempting to confuse the legal issues at bar is without merit. Plaintiffs are aggressively moving toward obtaining evidence that will conclusively show that Plaintiffs' injuries are a result of the negligence of the United States and that will overcome the government's immunity defense.

Conforti moved quickly to keep the Discovery from proceeding. He adroitly sidestepped any of the issues cited in the request for the Discovery to continue by filing a "Motion for Protective Order Concerning the First Request for Production." In this filing, he stated that "much of the information requested is duplicative of the Interrogatories already answered in this case."

This was not true, but it didn't seem to matter. He even claimed the request for Discovery was "unduly burdensome and constituted sheer harassment. of the defendant." He asked for "a protective order staying all discovery pending the Court's decision on the Motion to Dismiss and/or for Summary Judgement filed January 17, 1985.

In addition, Conforti provided typed copies of the "Protective Order" and an "Order for Dismissal" for Judge Ross Sterling's signature. The Protective Order was signed on May 28, 1986, and the Order for Dismissal was signed on August 21, 1986.

Ross notified Betty, Vickie, and I about the Order for Dismissal. In her letter dated October 21, 1985, she said:

I was sorry to hear that the Defendants Motion to Dismiss-the case was granted by Judge Ross Sterling. Sixty days has elapsed since that time, and by operation of law the case will be dismissed and unappealable on Tuesday, October 21, 1986. Although I had some people who were willing to work on the appeal, they could not do so without a Five Thousand Dollar ($5,000.00) retainer fee. Additionally, and unfortunately, we could not be assured that an appeal would be successful.

I know that everyone worked very hard and diligently to get the support and the information we needed. I'm afraid it came too late. I was never allowed to be substituted as the attorney of record for Peter Gersten. And as a result of that, I could only attempt to keep us in court.

Everyone had worked very hard to bring the case before a judge and the people; however, it never happened. The evidence was never heard. As a result, Betty and Vickie were very upset with Judge Ross Sterling because he didn't seem to care about how badly they had been injured or that the U.S. Government was involved through the helicopter units that participated during the event. Perhaps the proceedings would have turned out differently had Attorney Ross been added to the team a year or so earlier. We will never know. Nevertheless, *IT WAS OVER.*

As a postscript, Judge Ross Sterling died of cardiac arrest on January 14, 1988, in a Houston hospital where he was being treated for internal bleeding.

CHAPTER 38
The Pain Continues

The period of the legal proceedings was an especially bad time for Betty, Vickie, and Colby. They were suffering from the apparent effects of radiation exposure. Illness was the norm. They frequently went from feeling poorly to being incapacitated. Betty, for one, has been hospitalized more times than anyone else I know that has survived.

Betty and Vickie's survival during these dark days had been enhanced by their fight for justice. It gave them a goal, something they could wake up in the morning and fight for. Ending the legal case took away that hope for justice and made it more difficult for them to face the pain and misery that had become a way of life since the incident.

Instead of writing ad infinitum about each time one of them saw a doctor or was admitted to a hospital since the incident, I will end the account with summary statements about specific topics.

Betty Cash

I am proud to have known Betty. By working with her all of these years, she has become like a family member. I believe she honestly and clearly described the encounter as it happened that night. Since then, she has openly shared the account with government representatives, lawyers, doctors, the general public, and even some charlatans who tried to take advantage of her.

Several people have questioned how Betty could get along since she wasn't working. Some even started rumors that the U. S. Government was paying her expenses, which of course, was not true. The answer is that Betty was able to get along because of her family and friends. Her wonderful mother Pauline Collins spent her last years caring for Betty. But it was Betty's strong will and sheer determination to seek justice that kept her going during all of the bad times.

Betty has never worked a single day since the encounter because of poor health. Initially, she was fighting the effects of the radiation burns. During that period, her initial health state was well established. Later, when she developed nodules and skin cancers, it was already documented that she had been cancer-free prior to the incident. And in 1983, when she developed breast cancer and underwent a mastectomy, a review of her mammogram from January 1981 showed she had no indication of breast cancer at that time. In the years that followed, she developed bone marrow problems, bone brittleness, and joint problems. Intermixed with all of this, she had continual bouts with radiation dermatitis and inflammation of the heart sac. The visual evidence of her injuries was documented by photographs, and her many hospital stays, including a number of times in intensive care, was documented by her medical records.

Several of Betty's doctors went on record saying it was their professional medical opinion that she had been exposed to radiation. These are doctors who assess patients after radiation treatments, and they are familiar with the changes in the texture of the skin when it is irradiated. They all said Betty's skin was typical of that kind of exposure. In fact, they disapproved of the use of radiation for treatment of Betty's cancer because they didn't believe she could stand any more exposure. After the mastectomy, they had a difficult time finding enough undamaged skin for grafting. And the grafts they did were slow in healing.

In particular, Betty credits Dr. Bryan A. McClelland with keeping her alive and making her as comfortable as possible. In a letter documenting Betty's condition, Dr. McClelland said: "She has clear evidence of post-radiation dermatitis with chronic changes in her right hand...." And "Ms. Cash's radiation exposure could not possibly have been casual. There are no sources of radiation available to the public of the intensity required to cause her injuries."

Betty continues to suffer, but she is a fighter. On January 9, 1997, she suffered another heart attack and survived.

Vickie Landrum

I am equally proud to have known Vickie Landrum. Her quiet determination and her ability to suffer without complaining is inspiring to everyone who comes in contact with her.

The burns, sores, hair loss, eye damage, and partial paralysis of the face were a constant reminder of the encounter. With these visible medical conditions, she could no longer work in the food service industry to supplement the family income. At times she was too ill to care for her husband's needs, and this was especially upsetting to her when he became ill.

No doubt, however, her biggest concern was Colby. She worried daily about what was going to happen to him because of the radiation exposure. Right after the incident, doctors told her to keep a close eye on Colby's health because he could develop radiation sickness at any time, even as much as ten years later. And during the times when she was so ill that she was afraid she wouldn't survive, she worried about what would happen to Colby if something happened to her. Colby was her pride and joy. Perhaps that is what kept her going.

Vickie's strong faith in God was always like a shining light. During the encounter, she calmed Colby by telling him that Jesus would take care of them. Throughout all of the bad times after the event, she called on God for strength.

Initially Vickie treated her and Colby's skin condition with Johnson's Baby Oil. While it gave a little relief, of course, it was not a curing agent. So, Vickie shopped around and finally found an Avon Company skincare product that gave them some relief. It was fairly expensive, and in the amounts they needed during the months after the encounter, it was beyond her budget. We contacted Avon on Vickie's behalf, and they supplied her with two cases of the product, free of charge. Avon requested nothing in return for their kindness.

After surviving all of the pain and worry, Vickie was again devastated by the loss of her loving husband Earnest in 1996.

Colby Landrum

Everyone involved tried hard to protect Colby from exploitation. The trauma he experienced during the encounter was enough to cause him problems for several years to come. He didn't need additional problems.

We consulted with Vickie on how to proceed every time Colby faced a new problem. Before the incident, Colby was an out-going and active lad. He fished, played T-ball, and participated in other sports like a boy several years older than his age of seven. After the incident, he was sick, scared, and had bad nightmares. For the year after the encounter, he lost weight, had trouble with stomach pains and diarrhea. He experienced an abrupt increase in tooth decay, and his eyesight was impaired. When school started the following September, he wore jeans that were one size smaller than he had the previous school year.

Colby was always a good student, but after the encounter, he began having trouble with the subject just before the recess period. This meant he would have to forfeit his recess in order to improve his grades. This didn't make sense to Vickie, so she investigated why it was happening. As it turned out, Colby was afraid to go outside for fear the helicopters or the UFO would return. As soon as the teacher understood the problem and allowed him to be "safe," Colby's classwork improved again.

In helping Colby to understand what types of spacecraft NASA was building, I had Vickie bring him to the Johnson Space Center. At the time, my office was in Mission Operations Directorate Building number 4, the same building where the astronauts had their offices. When Vickie and Colby arrived, I met them at the door to the building. At that point, I found that Vickie had parked in a reserved parking spot, and so I asked her to move the car to another parking lot so she wouldn't receive a citation. She left Colby with me standing outside of the building. We were talking about what he might like to see — the astronaut trainers, Gemini Spacecraft, Apollo Spacecraft, and the Skylab — when a U.S. Coast Guard helicopter came in from Galveston Bay and headed straight at us en route to Ellington Field. Colby changed from a happy, excited boy to a very frightened one. He started trying to dash for a bush or other cover, and I had to hold on to him and calm him. His great fear of the helicopter was so upsetting that I will never forget it. After the helicopter was gone, we had a great day looking at real space stuff.

The reaction of other children was tough on Colby. At times he was teased and mocked for seeing a UFO and for being afraid. This was very hard for him to deal with. The peer pressure brought on by other children is very harsh. Just telling Colby to ignore it did not work. He became very depressed, and Vickie worried that something might happen to him. It was about that time that the Houston CBS-TV affiliate decided to do a news series on the incident. We were all hesitant to participate because other news people had come in, got their story, and left without helping in any way.

Since we were still trying to reach some of the helicopter pilots to solicit their help, we hesitantly agreed to participate. This series turned out differently. The newsperson involved was Mitch Duncan, a very popular TV figure. Not only was he a good reporter, but he also had a background in law. His series on the case was outstanding. It was fair and truthful, with no hype. Mitch was so impressed by the interviews that he offered to take Colby along with his own family to the Houston Livestock Show and Rodeo; After that, Colby was a star. He had been with Mitch Duncan, and none of the other children mocked him again. We all said: "God bless Mitch Duncan."

I am happy to report that the worries of the doctors were not realized. Colby did survive the sores, nodules, weird body hair growth, weight loss, stomach pains and did not develop leukemia. In May 1992, Kathy and I had the pleasure of attending his graduation from the High School in Dayton. He went on to a trade school and now has a good job and is very happy. Vickie is very proud of him. Now he worries about her like she worried about him in those dark days.

The Car

Betty said her 1980 Oldsmobile Cutlass was a joy to drive. She cared for it like a baby, keeping it washed and polished. However, it was never the same after the encounter. The plastic-covered padded dashboard still carries the imprint of Vickie's fingers where she had leaned forward, propping herself so she could look up at the huge object above the car. The plastic had been hot, and it molded to the shape of her fingers. When we looked at it later, Vickie's fingers did fit the imprint exactly.

After the event, Betty had a lot of trouble with the electrical system. The paint on the exterior dulled rapidly in spots, and the windshield shattered when a movie light shined on it. The damage made it look like it had been hit in several places with a ball bat.

Perhaps the most mysterious problem was with the steering wheel. In 1987, when one of Betty's relatives was driving Betty back and forth for medical treatments and errands, she said the steering wheel was causing a skin eruption on her hands. When they checked, they found the steering wheel was starting to crack and fall apart. Then within weeks, all of the non-metallic parts of the steering wheel separated and fell to the floor of the car, leaving only the reinforcing wire in place. When the Oldsmobile company people replaced the steering wheel, they said they had never experienced a failure like this. They had no explanation for it and had nothing in their files to compare it with.

The Helicopters

The Government's handling of the helicopter issue was one of "smoke and mirrors." They were evasive, obstructionistic, and at times, they lied; but they got by with it. However, it is my hope that one day the Government documentation will be released admitting to its participation in the encounter.

Some people questioned how helicopters could be seen at night, so that became an important part of our investigation. I didn't blindly accept the statements by Betty, Vickie, and another half dozen or so witnesses when they said they could clearly see helicopters in the night sky. I went out and duplicated their observations on 26 other nights. The main difference, of course, was that the UFO was not present and lighting the sky as it did on December 29, 1980. The weather on those 26 occasions ranged from hot and clear, to cold, damp, windy, and chilly. I found that the heavy moisture content of the Houston air reacts like little crystals that catch the light from the city, the moon, and from cars and reflect it in an airglow manner that leaves the sky very light on a lot of nights. A deep, dark night sky in the Houston area is unusual.

As an example of what I was able to observe, I watched CH-47 flights over my own backyard at 6:19 p.m., 6:41 p.m., and 7:19 p.m. on December 28, 1982. It was night, but the helicopters were clearly visible. I made another observation of a CH-47 at 7:05 p.m. on December 30, 1982. At that time, I was able to photograph the CH-47 without a flash attachment using my Canon camera and ASA 400 film. Hence, I personally verified that low flying helicopters could be seen at night, just as reported by the witnesses. Further, I proved that the U.S. Government and the Ellington CH-47 people had lied when they said that they never fly during Christmas week and never on a Monday night. In fact, it became a joke around my house when we would hear and/or see the CH-47s flying on a Monday night; we would say: "Those pilots don't know what night it is again; they aren't supposed to be up there."

During the intervening years, I have built a huge file of newspaper clippings about military helicopter activities since 1980. The content of this file has convinced me that military units do operate at strange times and in strange ways, very much akin to the activities reported on December 29, 1980. A few examples follow.

A *Wall Street Journal* article, dated October 21, 1980, described the activities of the American Nuclear Emergency Search Team (NEST). According to the article, the team gets called out on drills like they are the real thing, but you won't hear about it because it is not public information. It said: "The team has a small fleet of airplanes and helicopters, a collection of sophisticated radiation detectors.... a mobile communications center and a small workshop." This is just one of a number of special-purpose military units that could participate in an activity anywhere in the world.

The *San Francisco Chronicle* on September 16, 1982, described how Army and Air Force Special Operations Forces operate under a "joint command." The article said the Army had 3,600 Green Berets and a large number of Rangers as a part of their "unconventional" force. The Air Force has a Special Operations Wing that uses unmarked aircraft for dropping the units into hostile territory. And if that is not enough, the Navy Seals have similar capability. Frank Conforti, the Government lawyer, failed to find any of these units, essentially saying that they did not exist.

The *Houston Chronicle* tells about a crash of a CH-47 on a small island in Lake Michigan on July 10, 1983. The news release said the helicopter was from the 101st Airborne Division at Fort Campbell, Kentucky. Later, it was found it was from the 160th Task Force, called the "Night Stalkers." It says: The 160th Task Force maintains such secrecy that: o A colonel dressed down his underlings in a back room at the funeral home during the Mark Rielly (killed in the crash) wake, accusing them of divulging too much information. The family contends they know virtually nothing about the crash to this day.

One widow had to file a request under the Freedom of Information Act and pay $289 in copying fees to read the army's accident report and account of her husband's death. The father of another dead Task Force flier had to file a similar request and pay $154.95 for photographs of his son's UH-60 Black Hawk helicopter accident near the Fort Campbell airfield.

One bereaved family had to submit in writing their questions about their son's crash and wait from 24 hours to several weeks for *carefully worded Army replies*.

Major Larry R. Sloan, commander of their son's company declined to discuss the accident with his parents, *citing secrecy*.

Investigations of the unit's crashes are *kept secret even from personnel at the Army's Aviation Safety Center* at Fort Rucker, Alabama, the clearinghouse for all crash and safety information.

An article entitled "Secret Commandos" was published in the Houston Chronicle on December 8, 1984. It told about the special team which is an upgraded and expanded version of the Delta Force based at Fort Bragg, North Carolina. The team, it said "trains with *special operations forces of the Army, Navy and Air Force.*"

Further, it says: ".... team is so secret that the Pentagon will not publicly disclose its name, nor any details about its makeup or operations." For its reporting relationship, it stated: "The team comes under the direct authority of the Joint Chiefs of Staff." Speaking about secrecy, it said ".... teams able to operate virtually undetected."

From time-to-time large groups of military helicopters have staged raids on U.S. cities for "training purposes." Usually, these activities don't get in the newspapers or on television news. No one admits that it is going on and no records are made public. My file contains information about several such "raids." Recent ones took place in and around Houston, Texas, in March and again in October 1996.

In the March 1996 raid, citizens in the Copperfield Subdivision in West Houston reported a large number of helicopters and gunfire at night. One homeowner said the big CH-47 helicopters, with twin rotors, "were practically shaking the pictures off her walls." The aircraft were "right over her house, and the lowest she ever saw." Finally, a spokesperson from Fort Bragg, North Carolina, admitted the exercise went on from Sunday through Wednesday nights in Houston. During the daytime, the CH-47s were staying at Ellington Field, across town from the Copperfield Subdivision. Lt. Col. Ken McGraw at Fort Bragg said there was no secrecy; "the exercise just seemed too brief and minor to warrant a news release." He said: ".... the Army conducts such training periodically over U.S. cities so their crews can learn to navigate over well-lighted, built-up areas." This is the opposite of what we were told in 1981 when government representatives said helicopter pilots "never" fly around bright areas or bright objects because it affects their "night vision." McGraw also said that airplanes must obey Federal regulations calling for a minimum altitude of 1,000 feet; but "helicopters may go lower if the operation is conducted without hazard to persons or property on the surface." In other words, they do whatever they want to do.

The October 1996 exercise was exposed when an Army Special Operations Command helicopter crashed near Sugar Land on the southwest side of Houston, Texas, and one of the crew was taken to Hermann Hospital for treatment. A spokesperson said the exercises were scheduled from October 27 through November 6 and were generally done between 6 p.m. and 10 p.m. He said the forces were practicing with helicopters, small arms fire and explosives in vacant buildings near downtown, the Houston Ship Channel and Sugar Land; but he would not say where the helicopters were staying while they were in the Houston area. He admitted only that a large number of helicopters were used in the exercise and the one that crashed was an OH-6 single rotor aircraft. This unit was attached to the 160th Special Operations Aviation Regiment at Fort Campbell, Kentucky.

The style of these operations was very similar to the Huffman military exercises on December 29, 1980. Large numbers of helicopters were used, the exercises were without announcement, and secrecy was the order of the day. Very few Houstonians ever knew the events took place.

The End Or The Beginning

Betty, Vickie, and Colby were let down by the Government they loved and trusted. They were left to suffer alone, without help or information about their abusers. One may only assume that the stakes in the game that the Government was playing were so high that the lives of innocent victims didn't matter. This fact is further demonstrated by the dollars the Government spent to keep from helping. It was phenomenal.

The Government's style is contrasted with the open, caring, honest approach displayed by Betty, Vickie, and Colby. They were always concerned that a similar thing might happen to other innocent bystanders and that the medical community would not know how to help them either. Because of this, I promised Betty and Vickie that I would publish a "catalog of medical injuries caused by UFO encounters" and would work to bring these effects to the attention of the medical community. Since then, I have written several papers on the subject, lectured at various symposia, including two sponsored by a Non-Governmental Organization of the United Nations, volunteered to administer the MUFON Medical Committee, and in 1996, self-published the catalog of medical injuries. In addition, I promised Betty and Vickie I would document their case in a book. With the publication of this book, I have fulfilled that promise.

Betty, Vickie, Colby, and I will never consider the case closed until the perpetrators are exposed, and justice is served. This is just the beginning....

Addendum

Copies of the original sketches, letters, and reports are provided in this section.

DEPARTMENT OF THE ARMY
HEADQUARTERS III CORPS AND FORT HOOD
FORT HOOD, TEXAS 76544-50

REPLY TO
ATTENTION OF

January 24, 1986

Directorate of Information Management

SrA John T. Dressler
206-A 19th St.
Hickam AFB, Hawaii 96818

Dear SrA Dressler:

This responds to your Freedom of Information Act requests dated January 10, 1986 addressed to III U.S. Corps Headquarters, Office of History and January 16, 1986 addressed to 1st Cavalry Division Headquarters, Office of History.

There were fifteen CH-47 helicopters stationed at Fort Hood during the time frame referred to in your request. These helicopters did not participate in an unclassified exercise or deployment during the period which included December 28, 1980.

Fee for search and reproduction is waived.

Sincerely,

Barbara A. Weaver
Freedom of Information
Official

CF:
6th Cav (AFVM-B)

DEPARTMENT OF THE ARMY
HQ, III Corps & Fort Hood
Fort Hood, Texas 76544-5056

AFZF-IM-AR

OFFICIAL BUSINESS
PENALTY FOR PRIVATE USE, $300

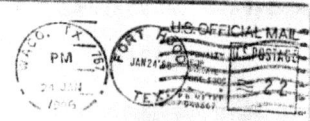

SrA John T. Dressler
206-A 19th St.
Hickam AFB, Hawaii 96818

FREEDOM OF
INFORMATION REQUEST

The Cash Landrum Incident

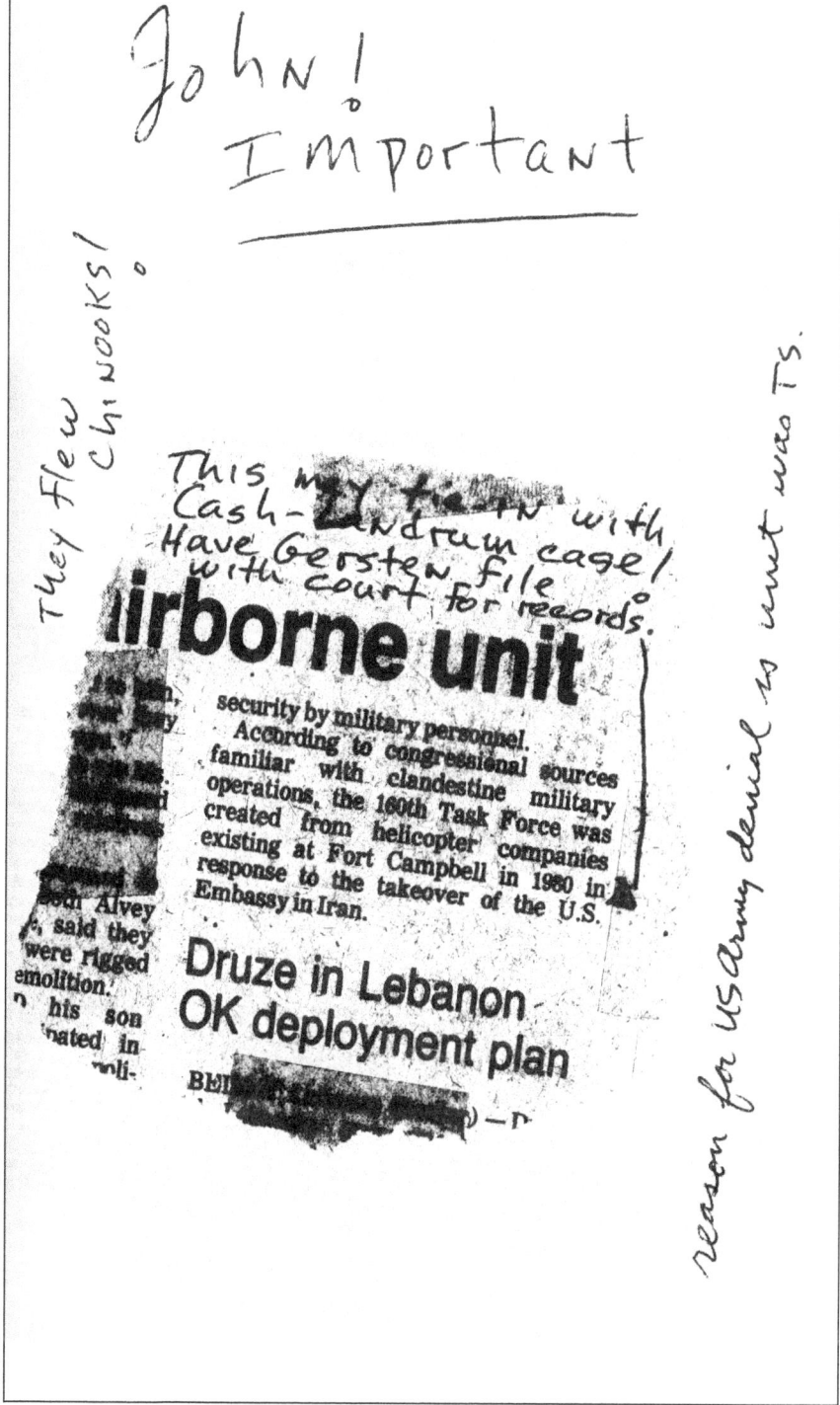

In Atom-Bomb Scare, Federal NEST Team Flies to the Rescue

* * *

Anti-Terrorism Squad Has Few Calls for Help Now, But the Potential Grows

By JOHN R. EMSHWILLER
Staff Reporter of THE WALL STREET JOURNAL

Oct 21, 1980 P. 1

LAS VEGAS—At 8 a.m. on a recent Sunday, retired Army Gen. Mahlon E. Gates was roused out of bed by a chilling phone call.

Terrorists, he was told, had built an atomic bomb and were threatening to detonate it "in a major U.S. oil field. If the federal government didn't meet their demands, the bomb would go off at 2 p.m. Friday.

Mr. Gates immediately began mobilizing his Nuclear Emergency Search Team, a U.S. government group of 200 specialists that was formed quietly in 1974. NEST's mission is to help prevent the detonation of a clandestine atomic bomb in this country.

By early Sunday evening Mr. Gates and his team were flying to the oil field. For the next few days they slogged past booby traps, finally reaching the weapon just before the deadline. They neutralized the bomb with a chemical explosion that flung plutonium debris over the nearby area.

You didn't hear about it because the entire operation was a drill, an elaborate simulation staged to measure NEST's ability to react to a real emergency. In the six years of its existence the team has rushed to three cities in response to such alerts, all of which turned out to be false alarms. "We hope we never find a bomb," says Mr. Gates, "but we have to anticipate that we might."

Knowledge, Materials Spreading

Concern over that prospect has intensified as both the material and the knowledge for making an atomic bomb spread around the world. "If a group really wanted a nuclear weapon it could get one," says Thomas B. Cochran, senior staff scientist for the Natural Resources Defense Council, a critic of the nation's nuclear program.

Others agree that the opportunities for acquiring atomic weapons are widening but doubt that a terrorist group would be willing to make the extensive effort. So far it appears that nobody other than a government has tried to make an atomic bomb.

"We aren't yet at the point that a bright lunatic can build one in his garage," says Brian Jenkins, a Rand Corp. specialist who has studied the potential for nuclear terrorism. But he adds that the drift of events is hardly comforting.

One alarming trend is the theft of nuclear material. Last year, for instance, an employe at a General Electric Co. plant in North Carolina stole 150 pounds of uranium intended for use in nuclear reactors and tried to ransom it for $100,000. The man was caught and convicted of extortion, GE says. So far nothing has been reported stolen that could be used in a weapon without further processing.

But some bomb-grade material is unaccounted for and could be missing. In 1965, officials of the since-closed uranium plant in Apollo, Pa., couldn't account for 200 pounds of highly enriched uranium, enough to make several bombs. No official seems to know what happened to that material, although some people believe it ended up in Israel for use in atomic weapons.

Tougher Safeguards Today

Apollo is the focus of a recent Nuclear Regulatory Commission report that describes how safeguards for nuclear material have been tightened. The report cites 134 ways that the Apollo plant, which met or exceeded all 1960s standards, fell short of current regulations. For example, previous rules allowed people to enter and leave storage areas without being searched.

These improvements haven't satisfied the critics. Nuclear facilities still have trouble accounting for all the weapons-grade material they handle. Over the years, thousands of pounds have been classified as "material unaccounted for." This may be nothing more than the material that is stuck in the maze of pipes and equipment that make up the nation's nuclear plants, but it's possible that some of it has been stolen.

Another point of criticism is that small quantities of bomb-grade uranium and plutonium may be shipped around the country without guards. The General Accounting Office warned last year that such shipments "may present an easy target for diversion by terrorists." The NRC says it would be too expensive to put guards on these shipments, which don't contain enough material to build a bomb. Nevertheless, the agency is reconsidering.

All of which leaves the critics unimpressed. "Before, we were just strolling along. Now we are jogging, but unless we start running as fast as we can we are going to lose this race," warns Theodore Taylor, once a leading bomb designer and for years a crusader for better safeguards over nuclear material.

Others worry that terrorists might be able to skip all the bother of building a bomb by simply stealing one from the government. The GAO has also studied this possibility, but most of its reports are secret. "About the only way I can put it," says Cliff Fowler, who managed the studies, "is that each time we looked we were concerned with what we found." But he believes the government has made "good" efforts to improve weak spots.

Please Turn to Page 19, Column 3

Continued From First Page

When all else fails, there is NEST. "We're for after the barn door is open," says one NEST official.

NEST was formed after an incident in Boston, when the city received an anonymous threat to set off an atomic bomb unless a $200,000 ransom was paid. The Federal Bureau of Investigation and the Atomic Energy Commission, the NRC's predecessor, decided to search the city. It turned out to be a nightmare—it took 48 hours just to collect the needed equipment, and there were difficulties getting planes to transport it.

Sixty Threats in 10 Years

More than 60 similar threats — most quickly dismissed as hoaxes or the work of crackpots—have been made against government agencies and businesses in the past decade. To avoid repetition of the Boston experience, the government has spent nearly $100 million to equip and operate NEST.

The team has a small fleet of airplanes and helicopters; a collection of sophisticated radiation detectors, some fitted into briefcases and camera bags that allow team members to make a search unobtrusively, and a mobile communications center and small workshop for making tools to dismantle a bomb—both of which fit into a jumbo-jet cargo pod.

NEST officials estimate the first team members can be on their way within two hours of an alert. Only a few members of the team, which includes such specialists as nuclear physicists and physicians, work full time. The others are "volunteer firemen," as one NEST official calls them, many of whom work for the federal government. Not surprisingly, many design and build nuclear weapons.

NEST occasionally gets called out on nuclear emergencies that aren't bomb hunts. They did some radiation testing at Three Mile Island last year during that plant's accident and helped look for the nuclear-powered Soviet satellite that fell in Canada in 1978.

Bomb Scare in Los Angeles

The team's searches for clandestine nuclear weapons have taken it to Los Angeles in 1975, Spokane, Wash., in 1976 and San Francisco this year. The latter two were brief episodes that required only a few NEST members. In the Los Angeles operation, though, the full team was mobilized when the chairman of Union Oil Co. received a threat that a nuclear explosion equal to 28,000 tons of TNT would be set off on company property unless a $1 million ransom was paid.

NEST searched Union Oil plants and offices for 2½ days; it even checked the chairman's home. The biggest discovery was a small piece of raw uranium that a company official kept in his desk as a souvenir.

Even if there had been a bomb, NEST might have had difficulty finding it. Radiation emitted by a weapon, which is NEST's only clue, can be blocked by many simple methods. So NEST works best in a small area.

"If you can cut it down to a few square blocks we have a chance. But if the message is to search Philadelphia, we might as well stay home," says Troy Wade, NEST's assistant director.

On 29 December 1980, Betty Cash, Vickie Landrum, and Colby Landrum were injured during a UFO close encounter near Huffman, Texas. During the incident, more than 20 military-type helicopters were seen by numerous witnesses, including the victims. During the extensive investigation of the case, government representatives stated for the legal record that none of their helicopters were at the scene of the incident. In addition, they stated that military helicopter bases were too far from the site for helicopters to get there due the the distance being beyond the helicopter's fuel range. Investigators found these statements to be untrue at the time. Now, once again, those same types of helicopters were in the Houston area and this time it is admitted in the following articles that the helicopters came all the way from Tennessee.

Hard landing by Army copter hurts 2

An Army Special Operations Command helicopter being used for urban military training "hard-landed" near Sugar Land Tuesday night, injuring the pilot.

Authorities said the helicopter landed about 7:15 p.m. near Flanigan Road, close to the intersection of U.S. 90A and Texas 6, just southwest of Sugar Land.

Police did not know what caused the accident, but an Army official said the aircraft had rolled over on its side after the hard landing.

Police said the pilot, 28, was flown by helicopter to Hermann Hospital, where he was listed in good condition Tuesday night.

Police reported that a second person was hurt, but that could not be confirmed Tuesday night.

The landing occurred on property owned by the Texas Department of Criminal Justice, police said. The Sugar Land Municipal Airport is less than a mile north of Flanigan Road, but it was not known if the airport was being used to stage the training exercise. The maneuvers were canceled for the rest of the night, police said.

The military operations are scheduled to be conducted in the Houston area until Nov. 6.

Houston Police Chief Sam Nuchia said in a press conference before the accident that the exercises began Monday, but his department had not notified citizens.

"This is a necessary Army practice," Nuchia said. "There's no danger to the public."

The forces will be practicing with helicopters, small arms fire and explosives in vacant buildings near downtown, the Houston Ship Channel and Sugar Land.

All of the test facilities are several blocks from residential areas, and police will block traffic into or near the sites during each exercise. Most are scheduled between 6 p.m. and 10 p.m.

Houston Chronicle, Wednesday, Oct. 30, 1996

Rumors fly as Army copters buzz city

By JOHN MAKEIG
Houston Chronicle

If you've been getting buzzed by Army helicopters and hearing what sounds like gunfire at odd hours, it's because Houston represents a unique environment for certain types of aerial maneuvers.

More exactly, the 11 MH-60 Blackhawk and OH-6 helicopters working around the Houston Ship Channel, the downtown area and Sugar Land are using Houston as a "MOUT site."

That's the Army acronym for "Military Operations in Urban Terrain."

Attached to the 160th Special Operations Aviation Regiment at Fort Campbell, Ky., the helicopters are practicing maneuvers through Wednesday.

The exercise started this week with 12 copters, but one OH-6 more or less crashed during a "hard landing" Tuesday evening near U.S. 90A and Texas 6 on Texas Department of Criminal Justice property near Sugar Land. The copter ended up on its side, slightly injuring at least one person aboard.

As Army spokesman Walt Sokalski explains it, the places where Special Operations forces are used are not always accessible and easy to use. In fact, Special Operations forces get sent to places often difficult for man and machine.

Houston has a varied landscape — from skyscrapers to open spaces to the Ship Channel — and many challenges for helicopter warriors.

Forget the Army posting notices about exercises so people can bring their families to watch the action.

"It's like when the circus comes to town. Everybody wants to see the elephants," Sokalski said. "It creates a greater hazard if we have a lot of people trying to watch the training."

Consequently, the Army or local police try to alert people in the immediate vicinity of the training two or three hours beforehand, but everybody else will hear hours later, if not the next day.

For instance, the current exercise started Monday, but the first public announcement came Tuesday — the same day the OH-6 took a tumble outside Sugar Land.

Doing things this way, unfortunately, has the side effect of generating rumors, and some people think the media are covering up for the culprits.

One caller to the Chronicle Wednesday demanded to know if there really are "3,000 soldiers shooting guns" around town.

Other callers suggested there was ample room to train around Fort Bragg, N.C., where, one woman insisted, the helicopters are based. "Who invited them here?" a caller wanted to know.

Sokalski declined to say where the Army helicopters are staying during the exercises, but four CH-47s used Ellington Field as a base when they were in Houston earlier for practice.

Houston Chronicle, Thursday, Oct. 31, 1996

TELECON / ~~CONFERENCE~~ / ~~TRIP~~ REPORT
(CROSS OUT NON-APPLICABLE TITLE)

PAGE 1 OF 1

VISIT
Vehicle Internal Systems Investigative Team

SUBJECT/PURPOSE: CASH/LANDRUM INVESTIGATION

DATE OF TELECON/CONF.: 5/26/82

C.W.O. W.F. Gustafson - U.S. Army Reserve Called

DISCUSSION/COMMENTS (INFO. OBTAINED, CONCLUSIONS)

His telephones are (work) 376-2950 & 376-2995
(home) 462-2318

1. He had done some checking with other members of the unit at Hooks Airport & suggested the following:

   a) There was a Quick React Force operating in Louisiana & Texas during the last year & a half. The last they heard of it was about 6 months ago, operating near Morgan City, La. He said they practiced "Iran type" raids, operating from a small carrier in the gulf of Mexico. Other times they haul in 5000 gal. fuel bladders for refueling. Their operation is secret & unannounced.

   b) USMC in New Orleans operates CH-46's

   c) Ken DeFore (HPD) lives in Dayton, Tx. He will visit L.L. Walker (Dayton policeman that saw the helicopters on 12/29/80.

PREPARED BY: [signature]   **DATE:** 5/26/80

P.O. Box 877 • Friendswood, Texas • 77546

The Cash Landrum Incident

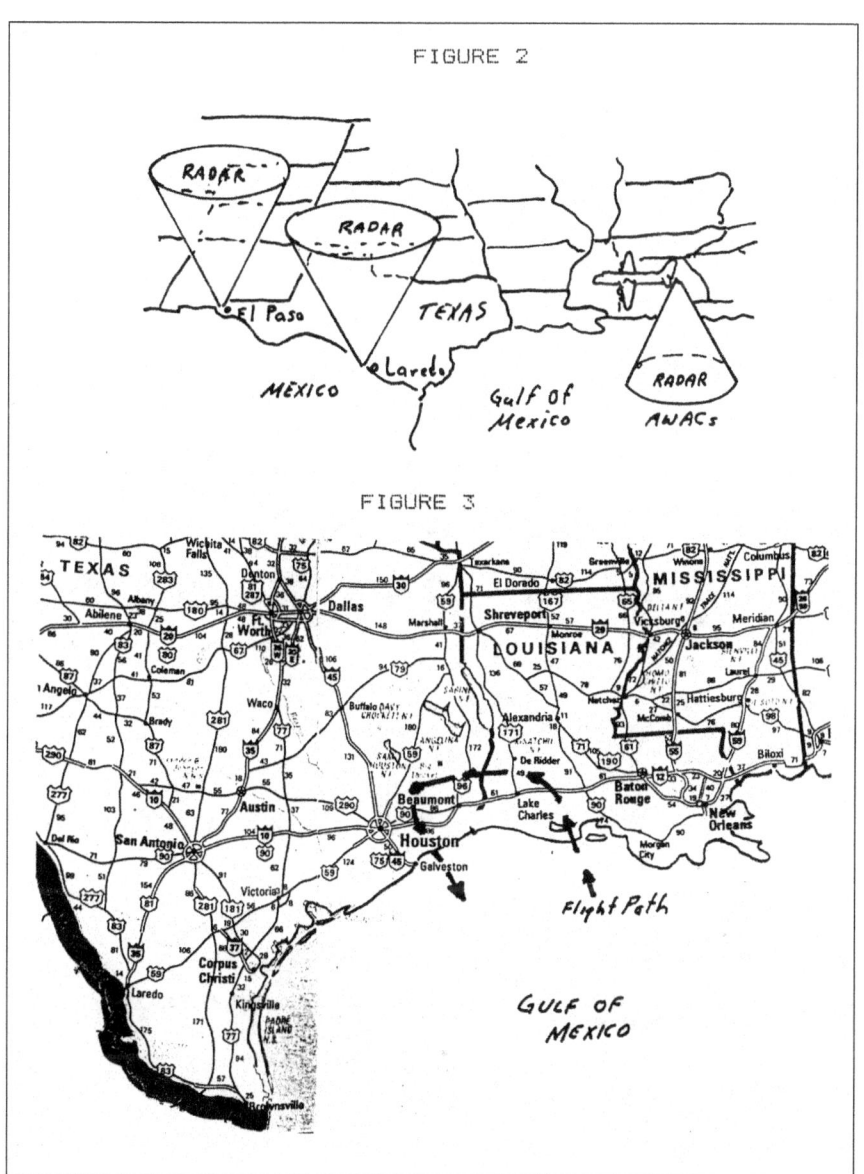

FIGURE 2

FIGURE 3

3848 Brighton Court
Alexandria, VA 22305

July 27, 1982

Inspector General
Department of the Army
Attention: FOIA Coordinator
The Pentagon
Washington, DC 20310

Dear Sir:

    Under terms of the U. S. Freedom of Information Act, I hereby request that you furnish me a copy of all U. S. Army records relating to the widely publicized Cash-Landrum incident of December 29, 1980, near Dayton, Texas, in which figured a flight of military helicopters.

    It is assumed that these records include both printed material and audio/video recordings of interviews; that they contain the incident's entire investigative case file developed by your investigator Lt. Col. George Sarran, USA; that they reveal the identity, mission purpose, and post-flight accounts of the helicopter crews involved; that they include all Serious Incident Reports (SIR) on the extent of any military operations associated with the incident; and that they include all records on the incident produced by the Army's Assistant Chief of Staff for Intelligence.

Yours sincerely,

Frederic L. Whiting

Copy furnished to:

  Chairman, Subcommittee on Government Information and Individual Rights, U. S. House of Representatives

DEPARTMENT OF THE ARMY
OFFICE OF THE INSPECTOR GENERAL
WASHINGTON, D.C. 20310

REPLY TO
ATTENTION OF:

DAIG-ZXF

19 AUG 1982

Mr. Frederic L. Whiting
3848 Brighton Court
Alexandria, VA 22305

Dear Mr. Whiting:

This responds to your letter of 27 July 1982 requesting a copy of a report of investigation. Your request was received in this office for reply on 6 August 1982.

Inspector general records are closely protected and controlled in order that the inspector general may effectively fulfill his responsibility as the confidential representative of his commander. It is my position that the records you have requested are exempt from mandatory release under the Freedom of Information Act, 5 USC 552(b)(5), (6) and (7) and paragraphs 2-12e, f and g, AR 340-17; however, I have considered your request and, as the Initial Denial Authority, have decided to furnish you the records with the following exceptions:

    a. Certain portions of the report that contain the opinions, conclusions and recommendations of the officer conducting the investigation. This material is considered to be intra-agency memoranda and, as such, is exempt from mandatory release under the Freedom of Information Act, 5 USC 552(b)(5) and (7) and paragraphs 2-12e and g, AR 340-17. In order for an inspector general to serve his commander effectively, he must be able to communicate frankly and fully without concern for public disclosure.

    b. Certain portions which contain material concerning other individuals, the release of which would be considered an unwarranted invasion of the individuals' privacy. This material is exempt from mandatory release under the Freedom of Information Act, 5 USC 552(b)(6) and (7) and paragraphs 2-12f and g, AR 340-17.

    c. Certain portions of the record did not originate within the U.S. Army inspector general system of records. These documents have been referred to the US Air Force inspector general, in accordance with Army Regulation 340-17. If you desire to appeal this initial denial, you should submit your appeal through this office to the Secretary of the Army, ATTN: General Counsel, Washington, DC 20310.

DAIG-AC   UFO Incident   Proposed Reply

SALL   DAIG-AC
ATTN: Mr. Maler   LTC Sarren/mr/51578

1. Attached is a draft reply to a Member of Congress concerning DAIG investigation of the UFO incident that allegedly occurred the evening of 29 December 1980.

2. Request you furnish this office your final response so we may close our file.

FOR THE INSPECTOR GENERAL:

1 Incl   SIGNED
as   ROBERT A. HARLESTON
                                   Colonel, IG
                                   Chief, Assistance Division

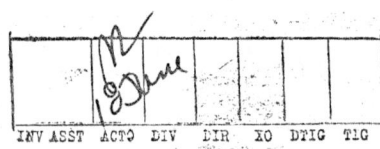

Dear Mr. Wyden:

This is in final response to your 16 February 1982 request to the Assistant Secretary of Defense for information concerning possible military involvement in an alleged UFO incident that occurred 29 December 1980 near the town of Dayton, Texas. An inspector general inquiry has been completed.

The allegation that Army, Army National Guard, or Army Reserve helicopters might have been involved on the evening of 29 December 1980 is not substantiated. Agencies queried included major Army commands, Army installations, test and evaluation agencies, National Guard and Reserve flying detachments, the Houston police helicopter unit, and others. Interviews were conducted with the victims and other persons thought to have information concerning the incident.

There was no evidence developed which supported the contention that Army helicopters were involved.

Sincerely,

Honorable Ron Wyden
House of Representatives
Washington, DC 20515

UFO Incident

26 Apr 82

MFR: Have made numerous inquiries as to the possibility that the aArmy may have been involved in the UFO incident in Dec 1980. All feedback has been negative. Will forward to the Houston area to inquire about helicopters that supposedly surrounded the "UFO" that caused severe burns to the people involved. Key players are John Schusaler, 713-483-2609, Dr. Niemtzow (a) 837-2140, Dr. Rank (Radiologist) 608-251-2371. COA: 1121 (Request for info). 26 Apr. No suspense.

G. C. SARRAN
LTC, SG

MFR: all memo to Div Chief   18 June

DAIG-AC

MEMORANDUM FOR DIVISION CHIEF

The allegation that Army, National Guard, or Army Reserve helicopters might have been involved in a UFO incident that occurred 29 December 1980

DISCUSSION:

This case was given to Army Congressional Liaison by AF Congressional Liaison after the AF could not determine any involvement. The three victims (two women and a young boy) clearly recall viewing some 23 helicopters orbiting around the object. Through the process of identifying silhouettes, some of the helicopters were determined to be twin rotors, or CH47s (Chinooks). Since the Army has the preponderance of troop and heavy equipment helicopters, the case was transferred to the DAIG for inquiry.

The DAIG inquiry focused exclusively on the question whether Army, Army National Guard or Army Reserve helicopters were involved in this incident. There was no effort to substantiate the existence of unidentified flying objects (UFOs), the events that happened that evening, or the medical problems that allegedly have occurred to the three victims.

Prior to visiting the site area, numerous phone calls were made to the different Army commands to request that records be checked to determine if any helicopters were flying at the approximate time and location of the reported incident. It is noted that the site of the incident is some 35 miles northeast of Houston, Texas, near Dayton, Texas. The reported time and date were between 2100 hrs and 2130 hrs, 29 December 1980 (Monday). There were no scheduled maneuvers in the area, and most Army units traditionally observe very limited operations on half day schedules.

Requests for assistance for any pertinent information were made to FORSCOM, Operation and Reserve Training Division, and program director for new systems; TRADOC, Operations and Training; Aviation Command, project manager for aviation systems; DARCOM-IG; TECOM; OTEA; DCSRIA; Fort Hood-IG; TCATA (at Fort Hood), and the Corpus Christi repair facility. Coordination was made with John Schussler, project director for manned flight operations with NASA, Major Dennis Haire local commander for eight Chinook Texas National Guard helicopters stationed at Ellington AFB, south of Houston; and CW4 Gustofson, senior AST for seven Army Reserve huey helicopters stationed at Tomball civilian airfield, northwest of Houston. Other coordination was made with Dr. Rank, M.D., Radiologist; and Dr. Niemtzow, M.D., USAF, Radiation Oncologist (specializing in radiation for cancer patients); both of whom had interest and knowledge in the case. After a period of time to thoroughly check flight records, all reports concerning any known helicopters flying in that general area were negative.

A trip was then made to the Texas site area to interview people with pertinent knowledge. John Schussler was interviewed. He had followed the case since February 1981 and was thoroughly conversant with all aspects of the case. Vicki Landrum (older of two women victims) was interviewed. She testified as to the events that evening. She was adamant that she and the other two victims (Betty Cash and her grandson) had counted approximately 23 helicopters flying around the object shortly after the object had ascended back in the sky. She related the medical disorders that have happened to each of the three victims; including sores on skin, hair falling out, blackened fingernails, constant diarrhea, loss of appetite, and diminished eyesight. The medical evidence of deterioration of health seems almost irrefutable, but was not a primary consideration in the DAIG inquiry.

Ms. Landrum related an experience she had in May 1981, some 5 months after the alleged incident. An Army National Guard helicopter (CH47) from Ellington AFB landed in the Dayton town square to be on static display for a local celebration. In a conversation with the aircraft pilot, CW3 Culberson, Mrs. Landrum heard him to say that he was flying the evening of the incident in response to an emergency by the Montgomery County Sheriff's Department. When pressed for more details, the pilot responded that he was prohibited from adding more information because of national security. After the interview with Ms. Landrum the DAIG investigator telephoned Ms. Betty Case in Alabama to corroborate or add any knowledge to the incident supplied by Ms. Landrum. Ms. Cash had moved to Alabama some months earlier so that her mother could take care of her because of the continued deterioration of her health allegedly caused by the incident. The 8 year old boy was not interviewed.

Next, a local Dayton policeman and his wife were interviewed. He recalled a conversation that he had with his wife at approximately 0040 hours, 30 December 1981, some 3 hours after the alleged sighting by the three victims, as they were returning from a visit to her parents' home. Some 8 miles from Dayton and within 5 miles of the earlier sighting, the policeman and his wife heard loud noises and noticed helicopters flying in groups of three in a "V" formation. They vividly remember discussing that some maneuvers must be going on nearby, the lateness of the hour (sometime between 0030 hrs and 0100 hrs), the helicopters were flying lower than normal (400 or 500 feet from ground level), they were twin rotors, and some of the helicopters periodically would turn on spotlights or landing lights which indicated they might be looking for something. Although the policeman discussed his experience with others at the office the next day neither he nor his wife could give any other names of people who might also have seen helicopters that evening.

A trip was made to Conroe, TX to interview the local sheriff. Neither he, his deputies, nor the dispatcher on duty the night in question could recall any emergency or any reason why helicopters might have been requested or flying.

Chief Warrant Officer Culberson, full time employee and maintenance officer for eight Texas National Guard helicopters at Ellington AFB stated that he remembers talking to Ms. Landrum while his aircraft was on static display in Dayton, TX. However, he emphatically stated that he was not flying that evening, he knows of no one who was flying, and his response to Ms. Landrum was simply that he had heard on the media that some helicopters responded to a request for assistance by the Montgomery County Sheriff's Department.

Major Haire, the National Guard detachment commander and CW3 Culberson's CO, stated that none of his aircraft were flying that evening. He further stated that it would be most unusual for any flying on a Monday evening. Virtually all flying in the unit is done on week-ends with occasional make-up flying done on Thursday evenings. Also, all flight missions must have his approval before the flight.

An interview was conducted with CW4 Gustofson, the senior full time administrator for the seven Army Reserve huey helicopters located at Tomball Airfield on the northern edge of Houston. He stated that none of his helicopters were flying that evening. He further stated that six of his seven helicopters had large painted red and white crosses which would have been clearly visible from the ground, even during reasonable darkness. He stated that if any of his helicopters were flying that evening, he would necessarily have been involved.

Additional inquiries were made to the captain of the Houston Police Department in charge of helicopter operations, the local FAA spokesman, and the civilian helicopter repair facility at Montgomery County Airport. No one had any knowledge of helicopters flying in the area.

Mr. John Schussler stated that another person had earlier told him that he saw helicopters flying at the approximate time of the incident. However, that person refused to be interviewed or otherwise cooperate in the inquiry.

Upon returning to Washington, DC, requests for information were made to the IG at the JFK Center, Special Operations, AF and Navy IGs, CIA, and the Bergstrom AFB JAG. In response to letters from Senators Tower, Bentsen, and Congressman Wilson, representatives of the JAG at Bergstrom AFB interviewed the two ladies and the boy in August 1981. The results of the interview were provided this office. Although the apparent purpose of the interview was to submit a claim against the government, the JAG office at Bergstrom AFB presently knows of no claim submitted by the victims or their lawyer. In summary, no one could provide pertinent information that might involve Army helicopters.

GENERAL COMMENT

Ms. Landrum and Ms. Cash were credible. The DAIG investigator felt

The policeman and his wife were also credible witnesses. There was no perception that anyone was trying to exaggerate the truth. All interviewees were extremely cooperative and eager to be helpful in any manner. Through the course of inquiry the DAIG investigating officer tried to concentrate on any reason or anyone or organization which might have been flying helicopters that particular evening in that general area. There was no evidence presented that would indicate that Army, National Guard, or Army Reserve helicopters were involved.

GEORGE C. SARRAN
Lieutenant Colonel, IG
Assistance Division

UFO

30 Mar ①

- Radiation Oncologist / at Galveston
- David Grant Travis AFB
- 1 yr ½ in Service
- Texas Air nat'l Gd
- 1976 - France
- consultant Fr gov't 3 or 4 yrs
- med injury - official prog
- GEPAN - Dr Alan Esterle director
- minister of Armed Forces
- set-up series of test; actual or hoax
- Blue Book, project - UFO didn't exist
- Robertson panel - important to study
- Fr gov't will intercept
- 1 yr earlier / same circumstances reported all over the world
- ~~John~~ Schussler - flt mgr of space shuttle
  grandmother ~~Betty~~ Vicki
  Cash Landrum
- known John for years
- Peter Rank, Dr. Radiology dept in Madison, WI
- Was interested - no public attention
  AF studies UFO
- helicopters / unusual
- Tyler, TX
- pilot - actually seeing object
- technology, we're not aware of
- Ionizing radiation
- hand was not injured

(2)

- identify pilot
- John - medical problems
- type of radiation, can't explain
- gamma radiation, through metal
- microwave rad?
- skin biopsies — undetrm'ed
- other cases:
- 30 or 40 cases
- Tyler TX — bright object / burn chest, head
- saw boys (Christian Scientists)
- diamond shape burn — afterwards on my chest
- John Fowler from Mass — similiar burns
- small junction marks / also on jeans
- med info
- records with a physician
- ch, lympro kind cancer, interferon, study
- research, ultrophysiology, immune cells
- cells sensitive to radiation, book out
- If gov't — 1yr to work with...
- Peter Rank —
- Cpt Richard Niemtzow (nimzow)

1 June

- Quick Reaction Force
  5000 lb fuel blatter
- NRO - nat'l Recon organ
- Morgan City, La
- Op + Tng - SF
- SFIG 2364905/0181

RICHARD C. NIEMTZOW, M.D.
166 CANNON DRIVE
TRAVIS A.F.B., CALIFORNIA 94535 U.S.A.
TELEPHONE:

Maj Dennis Hair
713 578 0907 ~~[redacted]~~

Maj Jack Reebles
plt 3762995
ofn 3762990   273 medevac
maint 3762022
      3764671

Major Jere Gatlan   7562366

- 5 ladies / little boy   ①
- Dec 29, 1980
- 9 at night
- hy 1485 / Lake Houston
- bright object
- water tank
- flame out bottom
- diamond shape look / lights
- 5-10 minutes
- bus woman, 6-10 minutes
- car was hot / hasn't run properly
- radian heat
- Helicopters followed
- lower 20 helicopters
- CH 47, no marking
- 2 kinds; of H
- others trailed back

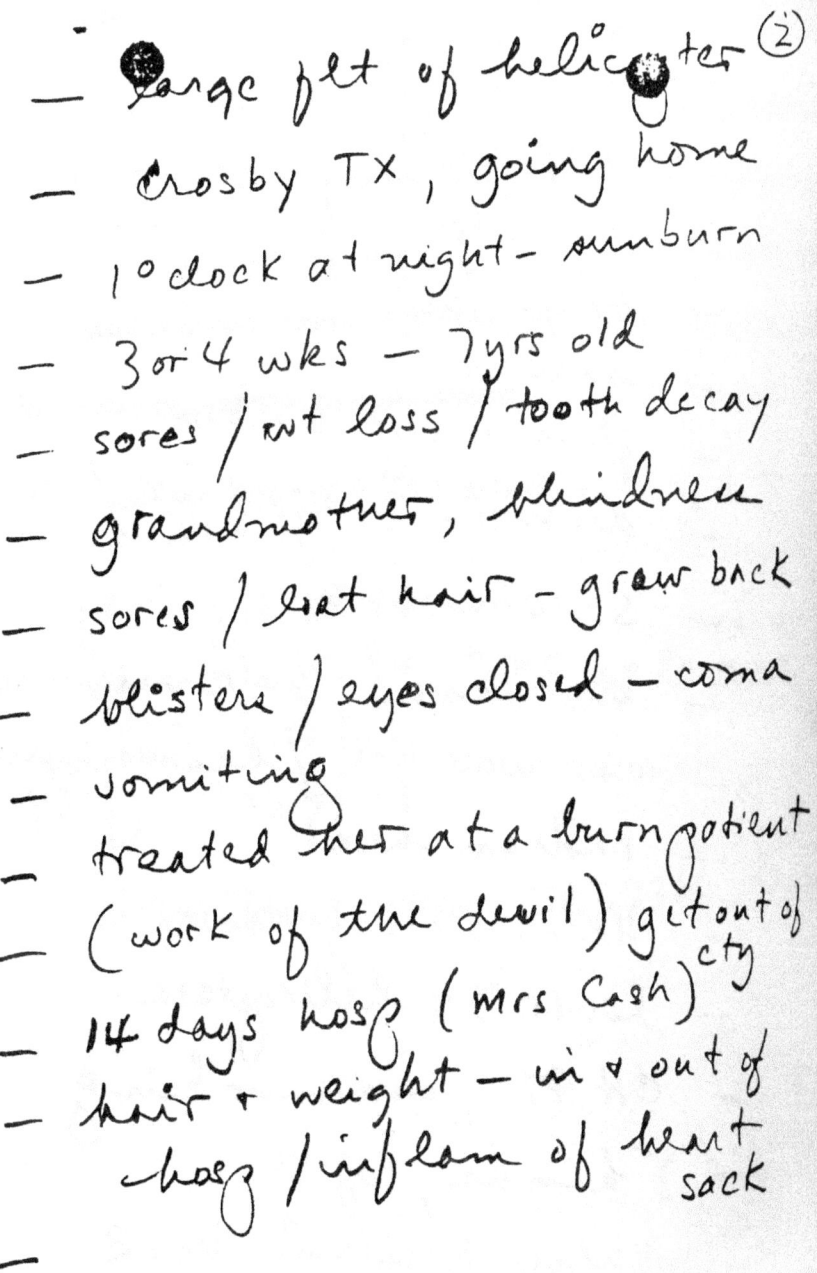

- large plt of helicopter ②
- Crosby TX, going home
- 1 o'clock at night — sunburn
- 3 or 4 wks — 7 yrs old
- sores / wt loss / tooth decay
- grandmother, blindness
- sores / lost hair — grew back
- blisters / eyes closed — coma
- vomiting
- treated her at a burn patient
- (work of the devil) get out of cty
- 14 days hosp (mrs Cash)
- hair + weight — in + out of hosp / inflam of heart sack
-

5 Apr

- DR Rauk
- radiation
- ionizing radiation, plus other radiation
- 2 physicans in US() in UFO
- long term med care
- Betty Cash - coronary art disease before
- chronic effects
- >10K, med expenses
- Betty Cash - Vicki Landrum (oldest) — grandchild Colby
- UFO - absolves US gov't
- OK - If John says so
- have not examined
- med record / hair loss / rad intoxication
- 25 yrs - Richard Hall - MD Ed + writer
  former asst dir 4418 39½ St Brentwood 20722
  AC 301
- MUFON - private organization (amateur)

6 Apr

- MG Chesney Murphy - Dg SG, AF
- Leo Sprinkle - hypnotist, Wyoming Univ Laramie
- Numerous flts to D.C. from Travis
- Dr Nientzow 837-2140

31 Mar

- LTC O'Connor 56512 DCSRTA
- 7379114/9602 Art Wood TCATA
- 2835171 Norm Atecker TECOM
- LTC Al Johnson 5882220/2926
- One, test supt
- AF liasion LTC Rockett 2891516/17
✓ - Col Fernandez - FORSCOM
- LTC Racine IG (Hood) 7377908/7209
- Maj Geo Jones 6803685 TRADOC
- Mr Don Reich 6804251 "
  ↳ LTC Thompson
- LTC McKiernan 2246/2044 Res tng div FORSCOM
✓ - Proj Mgr for avn systems - Avn Lee Dec 6937153
                                            Mr Lovett, prog
- Avn Div - LTC Hall 2709/3940   new system
  ↳ ferrying opn   → no records
- Reg'mts bt (tracking)(exercise)(ferry)(test)
- 6' prog, White sands
- buried in the area
- Fort Sam Houston, medavac
- Norm T Stecher TECOM 2835171/4544
- That's Incredible 213 4739641
  Stewart Schwatz

- Art Woods 7379114/9202 negative
- LTC Ellis Darcom IG

DEPARTMENT OF THE AIR FORCE
OFFICE OF THE SECRETARY

MEMORANDUM

MR for Army LL

SAF/LL contacted Mr. John Schuessler (McDonnell-Douglas) who is referenced in article. Mr. Schuessler has been in constant contact with individuals in article. Severe medical problems were confirmed and on going. Mr. Schuessler approached Army helicopter pilot with 136 Transportation Squadron, Ellington who allegedly bragged about participation in incident. Mr. Schuessler will provide pilot's name and other info on request. Very interesting conversation with Mr. Schuessler.

VIRGINIA A. LAMPLEY, Capt, USAF
Congressional Inquiry Division
Office of Legislative Liaison

```
                                                    2

— Peter Rank, Dr —
— AC 608-241.4611 (w)
  309 W. Wash Ave
  M. WI 53703
— Radiology consultants of Madison
— (h) 608-256 4139
— 16 May — marriage NJ   Phil
— meeting, 20 May
— med info, clearly suggest med sound
— Hamee, Dr   trained astronauts   2000
— boy — chinook  CH 47
— Dr Rosenthaul — new Dr
— 100 helicopters — Robert Grey airfield, came
     came in, for effect
— may be other witnesses                     Maine
— Thur 13 → Phil / 17-18 / 20 Thur / 22 May


— DR Niemtzow — AV 837-2140
```

DEPARTMENT OF THE AIR FORCE
HEADQUARTERS 67TH COMBAT SUPPORT GROUP (TAC)
BERGSTROM AIR FORCE BASE, TX 78743

4 June 1982

REPLY TO
ATTN OF: JAD/685-3784

SUBJECT: Landrum/Cash Interview

TO: DAIG-AC/Lt Col George Sarran
The Pentagon
Washington, D.C. 20310

1. Per our telephone conversation enclosed are copies of memos generated concerning an interview with Ms Vickie Landrum, Ms Betty Cash and Master Colby Landrum and Capt William J. Camp, Capt Terry Davis and Ms Pat Wolfe on 14 August 1981 regarding the alleged UFO sighting on 29 Dec 1980. Also attached are tapes of the interview (3 sides). This is a copy of the tape and need not be returned.

2. To date no claim has been received in this office.

3. This information has not been forwarded to any other agency. An attempt was made to forward this to Project Bluebook but, Maj Williamson, SAF/PACC (AV 227-1128) informed us that that office no longer existed and that the Air Force does not investigate UFO alleged sightings.

GERALD C. SHEA, Lt Col, USAF
Staff Judge Advocate

2 Atch
1. Memos
2. Tapes

Readiness is our Profession

United States Senate
WASHINGTON, D.C. 20510

COMMITTEES:
FINANCE
ENVIRONMENT AND PUBLIC WORKS
JOINT ECONOMIC

July 28, 1981

Ms. Betty Cash
209-48th Street
Fairfield, Alabama 35064

Dear Ms. Cash:

I have received your recent letter in which you describe the events that occurred on December 29, 1980.

Upon receipt of your letter, conversations were held with representatives of the Department of Defense. As a result of those conversations, it was suggested that you contact the Judge Advocate Claims Officer at Bergstrom Air Force Base, Austin, Texas, to file an official report and to submit a claim. I am advised that those officials have been made aware of your letter and the general situation which you outlined; they will be most willing to assist in any way possible.

Thank you for taking the time to write. I trust this will be helpful to you.

Sincerely,

Lloyd Bentsen

Lloyd Bentsen

> Betty Cash
> 8-18-81
> Vickie Landrum
> 8-18-81
>
> U. S. Air force — BC

NOTE: This is the drawing of the UFO by Betty Cash and attested to by Vickie Landrum.

United States Senate
WASHINGTON, D.C. 20510

September 4, 1981

Ms. Betty Cash
209 48th Street
Fairfield, Alabama 35064

Dear Ms. Cash:

Thank you for your recent correspondence.

I have been in contact with appropriate authorities and have been informed that you need to file a claim with the Base Staff Judge Advocate. His address is as follows:

        Base Staff Judge Advocate
        Attn: Claims Officer
        Bergstrom, Air Force Base 78743

I hope that this information will be helpful to you and if I can ever be of assistance to you in the future please do not hesitate to call on me.

Sincerely yours,

John Tower

JT/znd

CLAIM FOR DAMAGE, INJURY, OR DEATH	INSTRUCTIONS: Prepare in ink or typewriter. Please read carefully the instructions on the reverse side and supply information requested on both sides of this form. Use additional sheet(s) if necessary.	FORM APPROVED OMB NO. 43-R0597

1. SUBMIT TO: Base Staff Judge Advocate Attn: Claims Officer Bergstrom Air Force Base	2. NAME AND ADDRESS OF CLAIMANT (Number, street, city, State, and Zip Code) Betty Cash 209 48th Street Birmingham, Alabama 35064		
3. TYPE OF EMPLOYMENT ☐ MILITARY ☒ CIVILIAN	4. AGE 53	5. MARITAL STATUS Single	6. NAME AND ADDRESS OF SPOUSE, IF ANY (Number, street, city, State, and Zip Code) N/A

7. PLACE OF ACCIDENT (Give city or town and State; if outside city limits, indicate mileage or distance to nearest city or town) On FM Road 1485 between New Caney and Huffman Texas-7 miles out of New Caney Texas (see diagram)	8. DATE AND DAY OF ACCIDENT December 29 '80 Monday	9. TIME (A.M OR P.M) between 9:00PM- 9:30PM

10. AMOUNT OF CLAIM (in dollars)			
A. PROPERTY DAMAGE -0-	B. PERSONAL INJURY $10,000,000.00	C. WRONGFUL DEATH N/A	D. TOTAL $10,000,000.00

11. DESCRIPTION OF ACCIDENT (State below, in detail, all known facts and circumstances attending the damage, injury, or death, identifying persons and property involved and the cause thereof) At the above time and place claimant was driving a 1980 Olds automobile with two passengers, Vicki and Colby Landrum, when they observed an unconventional aerial object. According to the claimant, the object was extremely bright and appeared to have no distinct shape. The object was approximately 60-80 feet above the road and appeared to be the size of a 'city water tank". Furthermore the object was surrounded by a glow and appeared to have red and orange (continued on attached page

12. PROPERTY DAMAGE
NAME AND ADDRESS OF OWNER, IF OTHER THAN CLAIMANT (Number, street, city, State, and Zip Code)
N/A
BRIEFLY DESCRIBE KIND AND LOCATION OF PROPERTY AND NATURE AND EXTENT OF DAMAGE (See instructions on reverse side for method of substantiating claim)
N/A

13. PERSONAL INJURY
STATE NATURE AND EXTENT OF INJURY WHICH FORMS THE BASIS OF THIS CLAIM Claimant, within hours of the close encounter with the unidentified flying object, began experiencing the following symptons: extreme and prolonged headache, nausea, swollen neck, red blotches appearing on face and head and swollen earlobes and eyelids. Within the next few days her health deteriorated rapidly. (continued on attached page

14. WITNESSES	
NAME	ADDRESS (Number, street, city, State, and Zip Code)
Mrs. Vicki Landrum	506 West Clayton, Dayton, Texas 77535
Master Colby Landrum	506 West Clayton, Dayton, Texas 77535

I CERTIFY THAT THE AMOUNT OF CLAIM COVERS ONLY DAMAGES AND INJURIES CAUSED BY THE ACCIDENT ABOVE AND AGREE TO ACCEPT SAID AMOUNT IN FULL SATISFACTION AND FINAL SETTLEMENT OF THIS CLAIM

15. SIGNATURE OF CLAIMANT (This signature should be used in all future correspondence) Betty J. Cash	16. DATE OF CLAIM December 20, 1982

CIVIL PENALTY FOR PRESENTING FRAUDULENT CLAIM The claimant shall forfeit and pay to the United States the sum of $2,000, plus double the amount of damages sustained by the United States. (See R.S. §3490, 5438; 31 U.S.C. 231.)	CRIMINAL PENALTY FOR PRESENTING FRAUDULENT CLAIM OR MAKING FALSE STATEMENTS Fine of not more than $10,000 or imprisonment for not more than 5 years or both. (See 62 Stat. 698, 749; 18 U.S.C. 287, 1001.)

95-106

STANDARD FORM 95 (Rev. 6-78)
PRESCRIBED BY DEPT. OF JUSTICE
28 CFR 14.2

PRIVACY ACT NOTICE

This Notice is provided in accordance with the Privacy Act, 5 U.S.C. 552a(e)(3), and concerns the information requested in the letter to which this Notice is attached.

A. *Authority:* The requested information is solicited pursuant to one or more of the following: 5 U.S.C. 301, 28 U.S.C. 501 et seq., 28 U.S.C. 2671 et seq., 28 C.F.R. 14.3.

B. *Principal Purpose:* The information requested is to be used in evaluating claims.

C. *Routine Use:* See the Notices of Systems of Records for the agency to whom you are submitting this form for this information.

D. *Effect of Failure to Respond:* Disclosure is voluntary. However, failure to supply the requested information or to execute the form may render your claim "invalid".

INSTRUCTIONS

Complete all items—Insert the word NONE where applicable

Claims for damage to or for loss or destruction of property, or for personal injury, must be signed by the owner of the property damaged or lost or the injured person. If, by reason of death, other disability or for reasons deemed satisfactory by the Government, the foregoing requirement cannot be fulfilled, the claim may be filed by a duly authorized agent or other legal representative, provided evidence satisfactory to the Government is submitted with said claim establishing authority to act.

If claimant intends to file claim for both personal injury and property damage, claim for both must be shown in item 10 of this form. Separate claims for personal injury and property damage are not acceptable.

The amount claimed should be substantiated by competent evidence as follows:

(a) In support of claim for personal injury or death, the claimant should submit a written report by the attending physician, showing the nature and extent of injury, the nature and extent of treatment, the degree of permanent disability, if any, the prognosis, and the period of hospitalization, or incapacitation, attaching itemized bills for medical, hospital, or burial expenses actually incurred.

(b) In support of claims for damage to property which has been or can be economically repaired, the claimant should submit at least two itemized signed statements or estimates by reliable, disinterested concerns, or, if payment has been made, the itemized signed receipts evidencing payment.

(c) In support of claims for damage to property which is not economically reparable, or if the property is lost or destroyed, the claimant should submit statements as to the original cost of the property, the date of purchase, and the value of the property, both before and after the accident. Such statements should be by disinterested competent persons, preferably reputable dealers or officials familiar with the type of property damaged, or by two or more competitive bidders, and should be certified as being just and correct.

Any further instructions or information necessary in the preparation of your claim will be furnished, upon request, by the office indicated in item #1 on the reverse side.

(d) Failure to completely execute this form or to supply the requested material within two years from the date the allegations accrued may render your claim "invalid".

INSURANCE COVERAGE

In order that subrogation claims may be adjudicated, it is essential that the claimant provide the following information regarding the insurance coverage of his vehicle or property.

17. DO YOU CARRY ACCIDENT INSURANCE? ☐ YES, IF YES, GIVE NAME AND ADDRESS OF INSURANCE COMPANY (*Number, street, city, State, and Zip Code*) AND POLICY NUMBER. ☒ NO

18. HAVE YOU FILED CLAIM ON YOUR INSURANCE CARRIER IN THIS INSTANCE, AND IF SO, IS IT FULL COVERAGE OR DEDUCTIBLE?	19. IF DEDUCTIBLE, STATE AMOUNT
N/A	N/A

20. IF CLAIM HAS BEEN FILED WITH YOUR CARRIER, WHAT ACTION HAS YOUR INSURER TAKEN OR PROPOSES TO TAKE WITH REFERENCE TO YOUR CLAIM? (*It is necessary that you ascertain these facts*)

N/A

21. DO YOU CARRY PUBLIC LIABILITY AND PROPERTY DAMAGE INSURANCE? ☐ YES, IF YES, GIVE NAME AND ADDRESS OF INSURANCE CARRIER (*Number, street, city, State, and Zip Code*) ☒ NO

(11. Description of Accident cont.)

flames emanating from the bottom. Claimant stopped her automobile since the object was now blocking the road. Claimant and the two passengers proceeded to leave the vehicle and look at the object which was now hovering at treetop level approximately 135 feet from the people. Claimant experienced intense and excruciating heat which appeared to be caused by the object. After approximately five minutes claimant returned to her automobile and the object appeared to move further away. Claimant proceeded to drive along the road where approximately three miles away she observed what appeared to her to be approximately 23 military-type helicopters, several of which appeared to be double rotary type, in the general vicinity of the object. Finally there came a time when both the object and helicopters disappeared.

(13. Personal Injury cont.)

Her eyes closed completely and she could not see for several days. The red blotches became blisters of clear fluid. Claimant was unable to eat and continued to suffer nausea, vomiting and diarrhea. Four days after the incident, claimant entered Parkway General Hospital in Houston, Texas where she was treated and remained for twelve days. Furthermore claimant suffered severe loss of hair and loss of fingernails. After being discharged, she continued to suffer swellings, headaches and loss of appetite. A little over a week later she returned to the hospital for an additional fifteen days of treatment. During the following year claimant continued to have medical problems including, but not limited to, recurring blisters, constant headache and back pains, weight loss, chronic diarrhea and bowel problems. Claimant experiences intense pain when she takes a hot bath and must take cold baths. In October, November and December of 1981 Claimant was treated at the Lloyd Noland Foundation, Inc. (Hospital & Clinic) Fairfield, Alabama for various ailments associated with her encounter with the object. As of the date of this claim, claimant remains constantly tired continues to experience recurring headaches, chronic diarrhea and associated bowel problems.

SUMMARY OF MEDICAL EFFECTS: Erythema, acute photophthalmia, impaired vision, dystrophic changes in the nails, stomach pains, vomiting, diarrhea, anorexia, loss of energy, lethargy, scarring and loss of pigmentation, excessive hair loss, and hair regrowth of a different texture.

DIAGNOSIS: Radiation damage confined to the skin and the immediate subcutaneous area. The type and dosage of radiation is unknown-possible radiodermatitis secondary to ionizing radiation.

DEGREE OF PERMANENT DISABILTY: Unknown at this time

PROGNOSTS: Unknown at this time

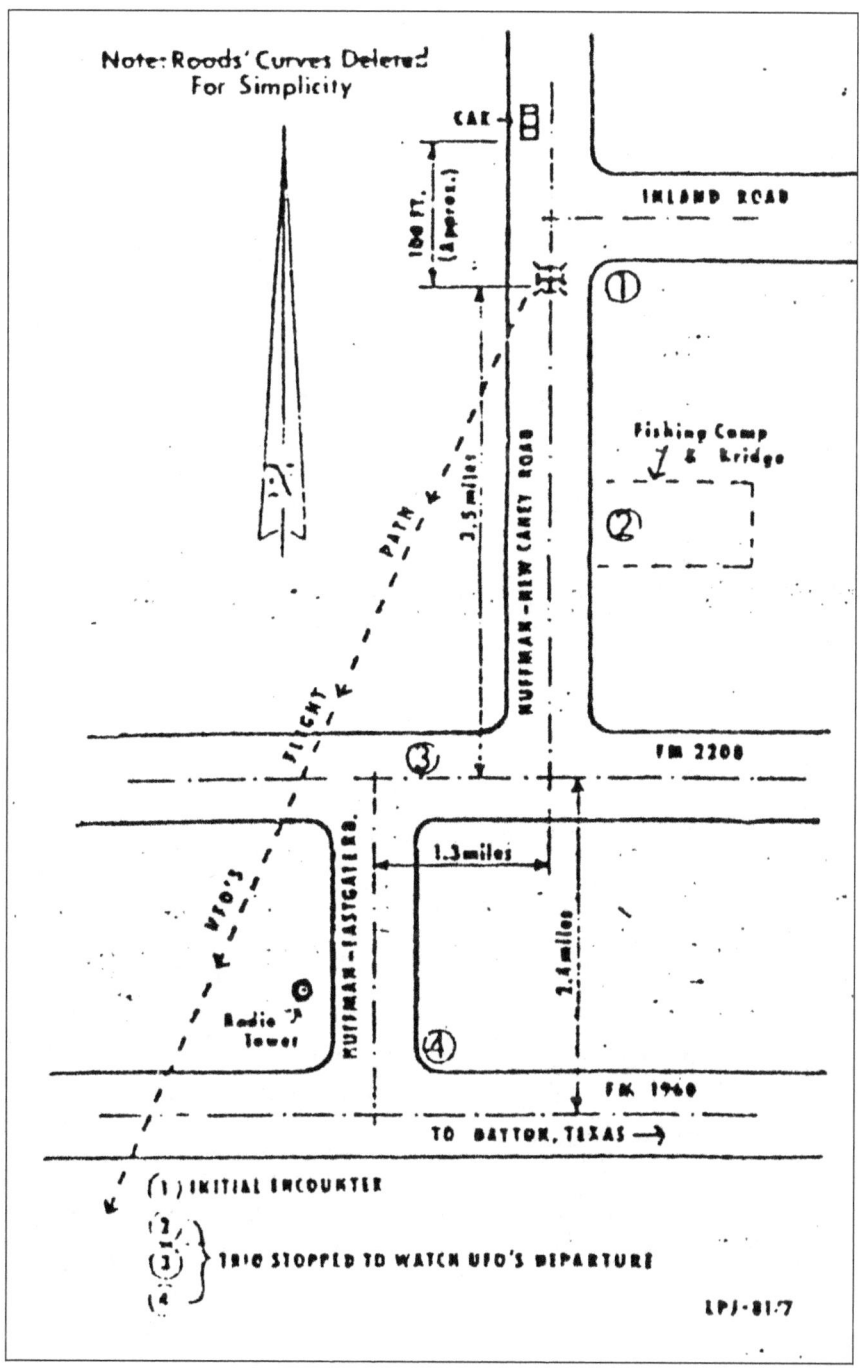

CLAIM FOR DAMAGE, INJURY, OR DEATH	INSTRUCTIONS: Prepare in ink or typewriter. Please read carefully the instructions on the reverse side and supply information requested on both sides of this form. Use additional sheet(s) if necessary.	FORM APPROVED OMB NO. 43–R0597
1. SUBMIT TO: Base Staff Judge Advocate Attn: Claims Officer Bergstrom, Air Force Base 78743	2. NAME AND ADDRESS OF CLAIMANT (Number, street, city, State, and Zip Code) Mrs. Vicki Landrum 506 West Clayton Dayton, Texas 77535	

3. TYPE OF EMPLOYMENT ☐ MILITARY ☒ CIVILIAN	4. AGE 59	5. MARITAL STATUS married	6. NAME AND ADDRESS OF SPOUSE, IF ANY (Number, street, city, State, and Zip Code) Earnest Landrum 506 West Clayton Dayton, Texas 77535

7. PLACE OF ACCIDENT (Give city or town and State; if outside city limits, indicate mileage or distance to nearest city or town) On FM Road 1485 between New Caney and Huffman Texas-7 miles outside New Caney Texas (see diagram)	8. DATE AND DAY OF ACCIDENT Dec. 29, '80-Mon.	9. TIME (A.M OR P.M) between 9:00PM-9:30PM

10.	AMOUNT OF CLAIM (in dollars)		
A. PROPERTY DAMAGE –0–	B. PERSONAL INJURY $5,000,000.00	C. WRONGFUL DEATH N/A	D. TOTAL $5,000,000.00

11. DESCRIPTION OF ACCIDENT (State below, in detail, all known facts and circumstances attending the damage, injury, or death, identifying persons and property involved and the cause thereof) At the above time and place claimant was a passenger, along with her grandson Colby, in an automobile driven by Ms. Betty Cash when they observed an unconventional aerial object. According to the claimant, the object was oblong with a rounded top and a point at the bottom and extremely bright. The object was 60-80 feet above the road and appeared to be the size of a 'city water tank'. Furthermore, the object was surrounded by a glow and appeared to have (continued on attached page)

12.	PROPERTY DAMAGE
NAME AND ADDRESS OF OWNER, IF OTHER THAN CLAIMANT (Number, street, city, State, and Zip Code)	N/A
BRIEFLY DESCRIBE KIND AND LOCATION OF PROPERTY AND NATURE AND EXTENT OF DAMAGE (See instructions on reverse side for method of substantiating claim)	N/A

13.	PERSONAL INJURY
STATE NATURE AND EXTENT OF INJURY WHICH FORMS THE BASIS OF THIS CLAIM	Claimant, within hours of the close encounter with the unidentified flying object, began experiencing the following symptons: stomach aches, loss of hair, indentations along her fingernails, burns of the face, swelling, redness and burning in her eyelids, diarrhea and nausea. During the next two years (continued on attached page)

14.	WITNESSES	
NAME	ADDRESS (Number, street, city, State, and Zip Code)	
Ms. Betty Cash	209 48th Street, Birmingham, Alabam 35064	
Master Colby Landrum	506 West Clayton, Dayton, Texas 77535	

I CERTIFY THAT THE AMOUNT OF CLAIM COVERS ONLY DAMAGES AND INJURIES CAUSED BY THE ACCIDENT ABOVE AND AGREE TO ACCEPT SAID AMOUNT IN FULL SATISFACTION AND FINAL SETTLEMENT OF THIS CLAIM

15. SIGNATURE OF CLAIMANT (This signature should be used in all future correspondence) Mrs. Vickie Landrum	16. DATE OF CLAIM December 20, 1982

CIVIL PENALTY FOR PRESENTING FRAUDULENT CLAIM	CRIMINAL PENALTY FOR PRESENTING FRAUDULENT CLAIM OR MAKING FALSE STATEMENTS
The claimant shall forfeit and pay to the United States the sum of $2,000, plus double the amount of damages sustained by the United States. (See R.S. §3490, 5438; 31 U.S.C. 231.)	Fine of not more than $10,000 or imprisonment for not more than 5 years or both. (See 62 Stat. 698, 749; 18 U.S.C. 287, 1001.)

95-106

STANDARD FORM 95 (Rev. 6–78)
PRESCRIBED BY DEPT. OF JUSTICE
28 CFR 14.2

PRIVACY ACT NOTICE

This Notice is provided in accordance with the Privacy Act, 5 U.S.C. 552a(e)(3), and concerns the information requested in the letter to which this Notice is attached.

A. *Authority:* The requested information is solicited pursuant to one or more of the following: 5 U.S.C. 301, 28 U.S.C. 501 *et seq.*, 28 U.S.C. 2671 *et seq.*, 28 C.F.R. 14.3.

B. *Principal Purpose:* The information requested is to be used in evaluating claims.
C. *Routine Use:* See the Notices of Systems of Records for the agency to whom you are submitting this form for this information.
D. *Effect of Failure to Respond:* Disclosure is voluntary. However, failure to supply the requested information or to execute the form may render your claim "invalid".

INSTRUCTIONS

Complete all items—Insert the word NONE where applicable

Claims for damage to or for loss or destruction of property, or for personal injury, must be signed by the owner of the property damaged or lost or the injured person. If, by reason of death, other disability or for reasons deemed satisfactory by the Government, the foregoing requirement cannot be fulfilled, the claim may be filed by a duly authorized agent or other legal representative, provided evidence satisfactory to the Government is submitted with said claim establishing authority to act.

If claimant intends to file claim for both personal injury and property damage, claim for both must be shown in item 10 of this form. Separate claims for personal injury and property damage are not acceptable.

The amount claimed should be substantiated by competent evidence as follows:

(a) In support of claim for personal injury or death, the claimant should submit a written report by the attending physician, showing the nature and extent of injury, the nature and extent of treatment, the degree of permanent disability, if any, the prognosis, and the period of hospitalization, or incapacitation, attaching itemized bills for medical, hospital, or burial expenses actually incurred.

(b) In support of claims for damage to property which has been or can be economically repaired, the claimant should submit at least two itemized signed statements or estimates by reliable, disinterested concerns, or, if payment has been made, the itemized signed receipts evidencing payment.

(c) In support of claims for damage to property which is not economically reparable, or if the property is lost or destroyed, the claimant should submit statements as to the original cost of the property, the date of purchase, and the value of the property, both before and after the accident. Such statements should be by disinterested competent persons, preferably reputable dealers or officials familiar with the type of property damaged, or by two or more competitive bidders, and should be certified as being just and correct.

Any further instructions or information necessary in the preparation of your claim will be furnished, upon request, by the office indicated in item #1 on the reverse side.

(d) Failure to completely execute this form or to supply the requested material within two years from the date the allegations accrued may render your claim "invalid".

INSURANCE COVERAGE

In order that subrogation claims may be adjudicated, it is essential that the claimant provide the following information regarding the insurance coverage of his vehicle or property.

17. DO YOU CARRY ACCIDENT INSURANCE? ☐ YES, IF YES, GIVE NAME AND ADDRESS OF INSURANCE COMPANY (*Number, street, city, State, and Zip Code*) AND POLICY NUMBER. ☒ NO

18. HAVE YOU FILED CLAIM ON YOUR INSURANCE CARRIER IN THIS INSTANCE, AND IF SO, IS IT FULL COVERAGE OR DEDUCTIBLE?	19. IF DEDUCTIBLE, STATE AMOUNT
N/A	N/A

20. IF CLAIM HAS BEEN FILED WITH YOUR CARRIER, WHAT ACTION HAS YOUR INSURER TAKEN OR PROPOSES TO TAKE WITH REFERENCE TO YOUR CLAIM? (*It is necessary that you ascertain these facts*)

N/A

21. DO YOU CARRY PUBLIC LIABILITY AND PROPERTY DAMAGE INSURANCE? ☐ YES, IF YES, GIVE NAME AND ADDRESS OF INSURANCE CARRIER (*Number, street, city, State, and Zip Code*) ☒ NO

(11. Description of Accident cont.)

red and orange flames emanating from the bottom. Ms. Cash proceeded to stop her vehicle since the object was blocking the road. Claimant, her grandson, and Ms. Cash all exited the vehicle to look at the object which was now hovering at treetop level approximately 135 feet from the people. Claimant experienced intense and excruciating heat which appeared to be caused by the object. After approximately three minutes claimant entered the automobile with her grandson leaving Ms. Cash outside. After a few moments Ms. Cash also returned and proceeded to drive down the road as the object appeared to move further away. After three miles claimant observed approximately 23 military-type helicopters, several of which appeared to be double rotary type, in the general vicinity of the object, which was still visible. Finally there came a time when both the object and helicopters disappeared.

(13. Personal Injury cont.)

has progressively gotten worse. She is presently suffering from a constriction in her left visual field-loss of vision in superior portion of the field. Furthermore, claimant continues to experience loss of hair, loss of weight, swelling in her legs and sores on her forehead.

SUMMARY OF MEDICAL EFFECTS: Photophthalmia, greatly diminished vision, stomach pains, diarrhea, anorexia, ulceration of the arms, resulting in scarring and loss of pigmentation, anychomadesis, hair loss, and hair regrowth of a different texture.

DIAGNOSIS: Radiation damage confined to the skin and the immediate subcutaneous area. The type and dosage of radiation is unknown-possible radiodermatitis secondary to ionizing radiation.

DEGREE OF PERMANENT DISABILITY: Unknown at this time

PROGNOSIS: Unknown at this time

CLAIM FOR DAMAGE, INJURY, OR DEATH	INSTRUCTIONS: Prepare in ink or typewriter. Please read carefully the instructions on the reverse side and supply information requested on both sides of this form. Use additional sheet(s) if necessary.		FORM APPROVED OMB NO. 43-R0597
1. SUBMIT TO: Base Staff Judge Advocate Attn: Claims Officer Bergstrom Air Force Base 78743	2. NAME AND ADDRESS OF CLAIMANT (Number, street, city, State, and Zip Code) Master Colby Landrum 506 West Clayton Dayton, Texas 77535		
3. TYPE OF EMPLOYMENT ☐ MILITARY ☐ CIVILIAN N/A	4. AGE 8	5. MARITAL STATUS single	6. NAME AND ADDRESS OF SPOUSE, IF ANY (Number, street, city, State, and Zip Code) N/A
7. PLACE OF ACCIDENT (Give city or town and State; if outside city limits, indicate mileage or distance to nearest city or town) On FM Road 1485 between New Caney and Huffman Texas-7 miles outside New Caney, Texas (see diagram)		8. DATE AND DAY OF ACCIDENT December 29,'80 Monday	9. TIME (A.M OR P.M) between 9:00PM-9:30PM

10. AMOUNT OF CLAIM (in dollars)			
A. PROPERTY DAMAGE N/A	B. PERSONAL INJURY $5,000,000.00	C. WRONGFUL DEATH N/A	D. TOTAL $5,000,000.00

11. DESCRIPTION OF ACCIDENT (State below, in detail, all known facts and circumstances attending the damage, injury, or death, identifying persons and property involved and the cause thereof) At the above time and place claimant was a passenger, along with his grandmother Vicki Landrum, in an automobile driven by Ms. Betty Cash when they observed an unconventional aerial object. According to the claimant, the object appeared extremely bright and diamond shaped. The object was approximately 60-80 feet above the road and appeared to be the size of a 'city water tank'. Furthermore,
(continued on attached page)

12. PROPERTY DAMAGE
NAME AND ADDRESS OF OWNER, IF OTHER THAN CLAIMANT (Number, street, city, State, and Zip Code)
N/A
BRIEFLY DESCRIBE KIND AND LOCATION OF PROPERTY AND NATURE AND EXTENT OF DAMAGE (See instructions on reverse side for method of substantiating claim)
N/A

13. PERSONAL INJURY
STATE NATURE AND EXTENT OF INJURY WHICH FORMS THE BASIS OF THIS CLAIM Claimant, during the close encounter with the unidentified flying object, became terrified and hysterical. Within the next several hours he experienced the following symptons: burning of the face which subsequently resulted in blisters, hair loss, redness and swelling of the eyes, irritation of the eyelids, stomach
(continued on attached page)

14. WITNESSES	
NAME	ADDRESS (Number, street, city, State, and Zip Code)
Mrs. Vicki Landrum Ms. Betty Cash	506 West Clayton, Dayton, Texas 77535 209 48th Street, Birmingham, Alabama 35064

I CERTIFY THAT THE AMOUNT OF CLAIM COVERS ONLY DAMAGES AND INJURIES CAUSED BY THE ACCIDENT ABOVE AND AGREE TO ACCEPT SAID AMOUNT IN FULL SATISFACTION AND FINAL SETTLEMENT OF THIS CLAIM

15. SIGNATURE OF CLAIMANT (This signature should be used in all future correspondence) Mrs. E.W. Landrum (as legal guardian)	16. DATE OF CLAIM December 20, 1982
CIVIL PENALTY FOR PRESENTING FRAUDULENT CLAIM The claimant shall forfeit and pay to the United States the sum of $2,000, plus double the amount of damages sustained by the United States. (See R.S. §3490, 5438; 31 U.S.C. 231.)	CRIMINAL PENALTY FOR PRESENTING FRAUDULENT CLAIM OR MAKING FALSE STATEMENTS Fine of not more than $10,000 or imprisonment for not more than 5 years or both. (See 62 Stat. 698, 749; 18 U.S.C. 287, 1001.)

95-106

STANDARD FORM 95 (Rev. 6-78)
PRESCRIBED BY DEPT. OF JUSTICE
28 CFR 14.2

PRIVACY ACT NOTICE

This Notice is provided in accordance with the Privacy Act, 5 U.S.C. 552a(e)(3), and concerns the information requested in the letter to which this Notice is attached.

A. *Authority:* The requested information is solicited pursuant to one or more of the following: 5 U.S.C. 301, 28 U.S.C. 501 *et seq.*, 28 U.S.C. 2671 *et seq.*, 28 C.F.R. 14.3.

B. *Principal Purpose:* The information requested is to be used in evaluating claims.

C. *Routine Use:* See the Notices of Systems of Records for the agency to whom you are submitting this form for this information.

D. *Effect of Failure to Respond:* Disclosure is voluntary. However, failure to supply the requested information or to execute the form may render your claim "invalid".

INSTRUCTIONS

Complete all items—Insert the word NONE where applicable

Claims for damage to or for loss or destruction of property, or for personal injury, must be signed by the owner of the property damaged or lost or the injured person. If, by reason of death, other disability or for reasons deemed satisfactory by the Government, the foregoing requirement cannot be fulfilled, the claim may be filed by a duly authorized agent or other legal representative, provided evidence satisfactory to the Government is submitted with said claim establishing authority to act.

If claimant intends to file claim for both personal injury and property damage, claim for both must be shown in item 10 of this form. Separate claims for personal injury and property damage are not acceptable.

The amount claimed should be substantiated by competent evidence as follows:

(a) In support of claim for personal injury or death, the claimant should submit a written report by the attending physician, showing the nature and extent of injury, the nature and extent of treatment, the degree of permanent disability, if any, the prognosis, and the period of hospitalization, or incapacitation, attaching itemized bills for medical, hospital, or burial expenses actually incurred.

(b) In support of claims for damage to property which has been or can be economically repaired, the claimant should submit at least two itemized signed statements or estimates by reliable, disinterested concerns, or, if payment has been made, the itemized signed receipts evidencing payment.

(c) In support of claims for damage to property which is not economically reparable, or if the property is lost or destroyed, the claimant should submit statements as to the original cost of the property, the date of purchase, and the value of the property, both before and after the accident. Such statements should be by disinterested competent persons, preferably reputable dealers or officials familiar with the type of property damaged, or by two or more competitive bidders, and should be certified as being just and correct.

Any further instructions or information necessary in the preparation of your claim will be furnished, upon request, by the office indicated in item #1 on the reverse side.

(d) Failure to completely execute this form or to supply the requested material within two years from the date the allegations accrued may render your claim "invalid".

INSURANCE COVERAGE

In order that subrogation claims may be adjudicated, it is essential that the claimant provide the following information regarding the insurance coverage of his vehicle or property.

17. DO YOU CARRY ACCIDENT INSURANCE? ☐ YES, IF YES, GIVE NAME AND ADDRESS OF INSURANCE COMPANY (*Number, street, city, State, and Zip Code*) AND POLICY NUMBER. ☒ NO

18. HAVE YOU FILED CLAIM ON YOUR INSURANCE CARRIER IN THIS INSTANCE, AND IF SO, IS IT FULL COVERAGE OR DEDUCTIBLE?	19. IF DEDUCTIBLE, STATE AMOUNT
N/A	N/A

20. IF CLAIM HAS BEEN FILED WITH YOUR CARRIER, WHAT ACTION HAS YOUR INSURER TAKEN OR PROPOSES TO TAKE WITH REFERENCE TO YOUR CLAIM? (*It is necessary that you ascertain these facts*)

N/A

21. DO YOU CARRY PUBLIC LIABILITY AND PROPERTY DAMAGE INSURANCE? ☐ YES, IF YES, GIVE NAME AND ADDRESS OF INSURANCE CARRIER (*Number, street, city, State, and Zip Code*) ☒ NO

(11. Description of Accident cont.)

the object was surrounded by a glow and appeared to have red and orange flames emanating from the bottom. Ms. Cash proceeded to stop her vehicle since the object was now blocking the road. Claimant, his grandmother and Ms. Cash all exited the vehicle to look at the object which was now hovering at treetop level approximately 135 feet from the people. Claimant became hysterical and experienced intense and excruciating heat which appeared to be caused by the object. After approximately three minutes claimant entered the automobile with his grandmother and approximately three minutes afterwards Ms. Cash returned. The three of them then proceeded down the road as the object appeared to move further away. After approximately three miles claimant observed approximately 23 military-type helicopters, several of which appeared to be double rotary-type, in the general vicinity of the object which was still visible. Finally there came a time when both the object and the helicopters disappeared.

(13. Personal Injury cont.)

aches, diarrhea and nausea. Claimant suffered from nightmares during the next several weeks and continues to display exterme aniety and fear at the sight of a helicopter. Claimant presently still suffers from recurring diarrhea, an inability to gain weight resulting from a continual loss of appetite, unusual thirst resulting in a large consumption of water, stomach pains after eating, and an increase in tooth cavities. His eyesight has progressively deteriorated to the point where he now wears glasses and still experiences unusual patches of hair growth on his back, neck and arms.

SUMMARY OF MEDICAL EFFECTS: Erythema, eyes swollen and watery, stomach pains, diarrhea, anorexia, weight loss, and an increase in tooth decay.

DIAGNOSIS: Radiation damage confined to the skin and the immediate subcutaneous area. The type of dosage and radiation is unknown-possible radiodermatitis secondary to ionizing radiation.

DEGREE OF PERMANENT DISABILITY: Unknown at this time

PROGNOSIS: Unknown at this time

DEPARTMENT OF THE AIR FORCE
HEADQUARTERS UNITED STATES AIR FORCE
WASHINGTON, D.C. 20324

2 SEP 1983

Mr. Peter A. Gersten
Gagliardi, Torres and Gersten
27 North Broadway
Tarrytown, NY 10591

Re: Appeal of Personal Injury Claims of Betty Cash, Vicki Landrum and Colby Landrum

Dear Mr. Gersten

The appeals of your clients' claims for personal injuries allegedly caused by an overflight of an unidentified flying object and unidentified helicopters on 29 December 1980 have been considered under 10 U.S.C. 2733 and are denied.

The reason for this decision is that the facts as alleged by the claimants fail to establish that their injuries were caused in any way by the United States Government or any of its agencies or instrumentalities. You should not consider the acceptance and subsequent denial of this claim as an admission of the truth of any facts alleged by your clients. Our investigation has revealed no evidence of involvement by any military personnel, equipment or aircraft in this alleged incident. The arguments you presented to establish liability of the government are not supported by any case or statutory law.

This is the final administrative action that can be taken on your clients' claims. This denial also satisfies the administrative filing requirements of the Federal Tort Claims Act. Based on this denial, your clients have the right to file suit against the government in an appropriate United States District Court not later than six months from the date of the mailing of this letter of denial.

Sincerely

Charles M. Stewart
CHARLES M. STEWART, Colonel, USAF
Director of Civil Law
Office of The Judge Advocate General

READY THEN   READY NOW

IN THE UNITED STATES DISTRICT COURT
FOR THE SOUTHERN DISTRICT OF TEXAS
HOUSTON DIVISION

CLERK, U.S. DISTRICT COURT
SOUTHERN DISTRICT OF TEXAS
FILED
JAN 3 1 1985
JESSE E. CLARK, CLERK
BY DEPUTY:

BETTY CASH, VICKI LANDRUM
and COLBY LANDRUM

v.

UNITED STATES OF AMERICA

CIVIL ACTION H-84-348

ORDER

Came on for consideration Plaintiff's unopposed motion for continuance of the trial setting in this case, and the Court having considered same, it is ORDERED that the motion is GRANTED.

It is further ORDERED that this case is reset for Docket Call on September 3, 1985, at 11:00 a.m., to be called for trial in its numerical order.

DONE at Houston, Texas, this 31st day of January, 1985.

United States District Judge

UNITED STATES DISTRICT COURT
SOUTHERN DISTRICT OF TEXAS
HOUSTON DIVISION

CLERK, U.S. DISTRICT COURT
SOUTHERN DISTRICT OF TEXAS
FILED
AUG 21 1986
JESSE E. CLARK, CLERK
BY DEPUTY:

BETTY CASH, et al §
 §
  Plaintiffs §
 §
vs. § Civil Action No. H-84-348
 §
UNITED STATES OF AMERICA §
 §
  Defendants §

ORDER OF DISMISSAL

CAME ON this day the Motion to Dismiss and/or for Summary Judgment filed by the United States and the Court, having considered the Motion and accompanying Memorandum, and the subsequent pleadings of the parties.

IT IS HEREBY ORDERED that the above noted cause of action is DISMISSED pursuant to Federal Rule of Civil Procedure Rule 12(b)(1), Rule 12(b)(6) and Rule 56.

DONE at Houston, Texas, this 21st day of August, 1986.

UNITED STATES DISTRICT JUDGE

RADIATION INJURIES FROM UFO

By Richard Hall

Preliminary Report

What promises to be one of the most significant physical evidence cases in modern UFO history occurred December 29, 1980, near Huffman, Texas, northeast of Houston. Two women and a young boy suffered various degrees of injury, largely attributable to radiation poisoning and radiant heat, after watching a luminous object hover low over the road ahead of their car. MUFON Deputy Director, John Schuessler, and members of Project VISIT are investigating. They are seeking to obtain the full set of medical records.

Betty Cash, 52, was driving her 1980 Cutlass Supreme from New Caney to Dayton, Texas, on Highway 1485 about 9:00 p.m. With her were a friend, Vicky Landrum, 60s, and Vicky's grandson Colby, 7. Suddenly a luminous, fiery-looking object descended to treetop level over the road ahead of them and they heard a beeping noise that persisted throughout the sighting. From its underside, flames (red-orange) were emitted toward the road periodically, with an audible "woosh." Betty stopped the car, afraid to drive beneath the object; they opened the car doors to stand beside the car and watch. The glow was brilliant, and they could feel strong heat and hear a loud roaring noise.

Colby became terrified and he and Vicky got back in the car, but Betty remained outside for a longer period of time. (Probably significantly, Betty's injuries were the most severe.) Finally, the object started to rise and move away to the right in a southwesterly direction with a large number of helicopters (20 or more) seemingly in pursuit. The evening was cool and the car heater had been turned on, but now the car was so hot that Betty turned on the air conditioner.

Later Betty dropped Vicky and Colby off at their house and drove home, feeling ill. She arrived home at 9:50 p.m. Numerous symptoms appeared almost immediately: swollen neck, head and facial blisters, swollen earlobes, and swollen eyelids. Her eyes closed completely and she could not see for several days. Four days later, unable to eat, and suffering nausea, vomiting, and diarrhea, Betty entered the hospital where she remained for 15 days. She also suffered severe loss of hair. After being discharged, she continued to suffer swellings, headaches, and lack of appetite. A little over a week later, she returned to the hospital for additional treatment. As of February 22, she remained constantly tired, headachy, and unable to work.

When they arrived home that night Vicky and Colby also felt ill; Colby's face was "sunburned" and he had eye problems, a condition that still persists to a mild degree. They spread large quantities of baby oil on their faces for three days. Both had stomach aches and diarrhea for several days. Vicky experienced some loss of hair and a sensation as if her scalp were "asleep." During the sighting, she had placed her left hand on top of the car, and the fingernails on that hand showed odd line-like indentations across their width.

Colby had nightmares for 2-3 weeks, and since has displayed extreme anxiety and fear at the sight of a helicopter. The large number of helicopters itself poses a mystery, since no obvious source of that many helicopters is known in the area, especially on short notice if they were pursuing the UFO, as they

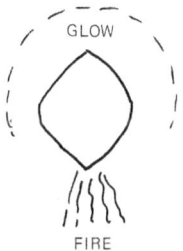

General appearance of Huffman, Texas, UFO

appeared to be. The date also was during the holiday season when military bases typically would be on "stand-down" with reduced personnel.

Although they remain to be fully documented, the medical symptoms suggest both radiation sickness and physical burns; both ultraviolet and infrared radiation may have been involved. Full details will be reported as soon as the investigation is completed, and the future health of the witnesses will be monitored.

CASH-LANDRUM RADIATION CASE
By John F. Schuessler

NOV 81

(Note: This follow-up report is based on a presentation to the Sept. 1981 CUFOS UFO Conference. The case was initially reported in the Apr. 1981 issue, No. 158.)

The problem of radiation sickness caused by UFOs is defined by these brief examples:

October 24, 1887: Venezuelan family exposed to a brightly lit unidentified flying object (UFO) and suffered burns, vomiting, hair loss, and extensive swelling.

May 20, 1967: Canadian prospector Stephen Michalak encountered a landed UFO and suffered burns, nausea, vomiting, swelling and an extended illness.

October 3, 1973: Missouri truck driver exposed to an extremely bright UFO, blinded for days, and had vision impairment for a year.

These and hundreds of similar incidents indicate that UFOs are seriously affecting people. How can these people be helped? What can we learn about UFOs by studying these human effects?

A small team of engineers, scientists, and medical specialists have formed Project VISIT (Vehicle Internal Systems Investigative Team), to be a clearinghouse for all UFO incidents involving medical injury or alleged entry into a UFO. VISIT members collect and analyze data on the physical effects of UFOs on people. This scientific and medical data is then examined to discover the probable mechanisms of the UFO.

The latest entry into the VISIT data base occurred on December 29, 1980, when three Texans encountered a UFO and suffered severe medical consequences. Betty Cash (51), Vickie Landrum (57), and Vickie's grandson Colby Landrum (7), were driving home to Dayton, Texas, on the Cleveland-Huffman road just north of Lake Houston. It was 9 o'clock at

Artist's Rendition of the Sighting
(By Kathy Schuessler)

night and the road was deserted. The first indication of something unusual was the presence of a very intense light several miles ahead just above the pine trees. Betty remarked about the unusual brightness, but temporarily lost sight of it due to the many trees along the road.

Suddenly, hovering over the road only a short distance ahead was an enormous diamond shaped object. "It was like a diamond of fire," Vickie said. The glow was so intense they could barely stand to look at it. Vickie at first thought it was the fulfillment of biblical prophecy and expected Jesus to come out of the fire in the sky.

In addition to lighting the whole area like daytime, the UFO periodically belched flames downward. Fearing they would be burned alive Betty stopped the 1980 Oldsmobile Cutlass without leaving the road. They all got out of the car to get a better look at the UFO. Colby was terrified and dove back into the car, begging his grandma to get back in, too. Vickie did and comforted Colby.

Betty stood momentarily by the driver's door and then walked forward to the front of the car. After much pleading by Vickie, Betty finally returned to the car. The door handle was so hot she used her leather coat as a hotpad to open the door. Although the winter night air had been about 40°F, the heat from the UFO caused the witnesses to sweat and feel so uncomfortable that they turned on the car's air conditioner.

Each time the object would shoot flames downward it would rise. As the flames stopped it would drop in altitude. The intense glow, however, never changed. In addition, the threesome heard an irregular beeping sound throughout the sighting.

(continued on next page)

Radiation, Continued

Chinook. Another was identified as being similar to the Bell Huey model, but not positively identified as such.

Each of the witnesses not only identified the shape and main characteristics of the Chinook, they also pointed out details such as the wheels, lighting pattern, and sounds.

Contact with the Houston International Airport FAA representative provided the following:

- 350-400 helicopters operate commercially in the Houston area.
- All are single rotor types (no Chinooks).
- Helicopter traffic flies Visual Flight Rules (VFR), consequently they do not contact the tower.
- Beyond 15 miles from the airport they must stay below 1,800 feet.
- The Houston radar is limited to 2,000-2,200 feet around Lake Houston due to the location of antenna.

Contact with military installations was of little help. Fort Polk, Fort Hood, Dallas Naval Air Station, and England AFB stated they did not fly into the Houston area that evening. The unit operating out of Ellington AFB in Houston had landed before the sighting time. Robert Gray Field had 100 helicopters come in from the field at one time "for effect," but claimed to have avoided the Houston area. Hence, no one claims the helicopters that filled the Huffman area sky that winter night.

Conclusion

This incident clearly points up several serious conditions. First, when a person is involved in a close encounter with a UFO they find it nearly impossible to obtain immediate assistance. The police, newspapers, and even doctors receive their plea for help with tongue in cheek. The doctors, being unprepared for a bizarre account like Betty's, spend a lot of time trying to determine what is wrong, as a standard treatment method has never been defined.

Second, military organizations could better serve the citizens of the United States if they were prepared to relate the nature of objects such as the one at Huffman and others where public safety is at stake. Betty and Vickie have never said the Huffman UFO was a flying saucer with little green men. They believe it was a government-sponsored operation of some kind. Others that saw and heard the helicopters that evening have the same feeling.

Third, UFO organizations usually do not cooperate to the fullest to help the witnesses. The Huffman incident is an exception. The Mutual UFO Network of Seguin, Texas, the Center for UFO Studies of Evanston, Illinois, and the Aerial Phenomena Research Organization of Tucson, Arizona, all cooperated in a responsible manner to assist the Houston-based Project VISIT to conduct the investigation by providing consultants, recommendations, and data pertaining to similar cases. Such cooperation is in the best interest of all parties involved.

The investigation continues. The future health state of Betty, Vickie, and Colby is yet to be determined. However, several radiation specialists have given freely of their time and talents to establish a program of rehabilitation and care. Full treatment is still lacking because the data on the source of the problem, the UFO, is still not available.

Project VISIT members are available on call for consultation. The address of VISIT is Post Office Box 877, Friendswood, Texas 77546.

(Photographs provided by John Schuessler)

Distinctive Silhouette of CH-47 Helicopter

Vehicle Internal Systems Investigative Team

Cash/Landrum/Landrum/
McConald/McDonald
12/29/80 Lake Houston Case
2/14/81

Page 1
D. Kissinger

Meeting 3/23/81
Kissinger/McDonald/McDonald

Jerry McDonald 23 years old
Glenda McDonald 22 years old
Baby girl approximately one year old

Sighted object directly overhead near his Trailer home. He saw it on a monday after Christmas between 7:00 and 8:00 P.M..(12/29) The object appeared over some 40 foot trees infront of his house and moved overhead off toward the Dayton high School. The object was triangular, Brilliant red light in the middle, white and blue lights on the end, brilliant cutting toarch white light shinning out from the aft end. The craft made a rumbling sound and moved approximately 3 miles/hour. The object did not have a tail or lights like an aircraft. Jerry looked at it for 2 to 3 minutes. It passed at an altitude of 130feet. The rumbling sounded like a jet engine when starting up. Jerry was in the yard to fix a water main. He was also going to take the trash out so he knows it was monday.

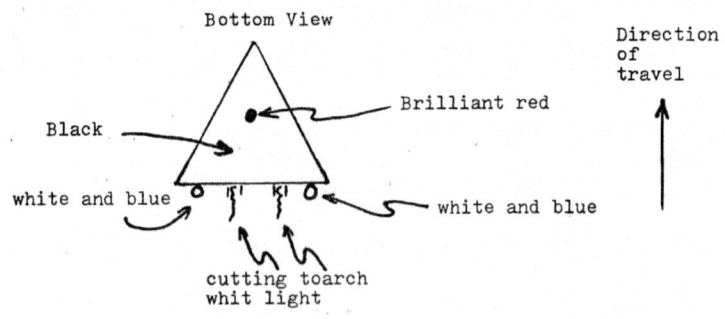

P.O. Box 877 • Friendswood, Texas • 77546

Meeting 3/23/81　　　　　　　　　　　　　　　　　　　page 2
Kissinger/McDonald/McDonald　　　　　　　　　　　　D: Kissinger

Jerry heard it before he saw it. John had flu and diarrhea beginning and the wednesday after the incident. He spent the day in bed. Glenda spent the day in bed with Jerry on thursday, both quite sick. The baby did not get sick. Jerry was hospitalized Feb 14 with an air pocket in his lung. It was treated with drugs. Jerry had 103 fever and headaches that wednesday. The fingernails showed no signs like those seen on Vickie and Betty.

Feb 14,1981 Glenda was coming from her mothers house at about 7:00P.M. Still light enough to see without lights. Stoped at Stop sign, saw tw͞o objects and got out of the car.

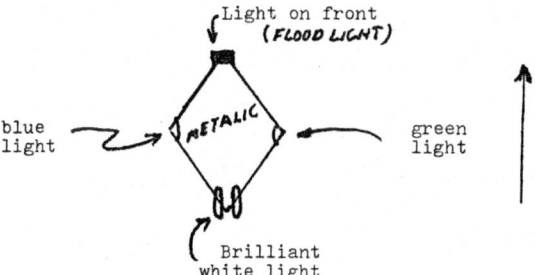

Made a humming noise. Moved approximately 10 miles/hr. Watched it for 15 minutes. Facinated Glenda to look at the two objects, They looked metalic. Was no fire coming out the back. Front looked like head lights. They looked like they were playing a game following each other. Jerry did not recognize the sketch Don made from Vicky's discription. Jerry estimated his object to be about 40 ft wide. Glenda's object was about 30 ft. Glenda's object was longer than wide. No problems with either car afterward. Jerry added that it sounded like the GoodYear blimp but it was the wrong shape.

Meeting 3/23/81 Page 3
Kissinger/McDonald/McDonald D. Kissinger

Jerry was sick off and on from the beginning of January until his hospitalization Feb 14.

The inference can be made that Jerry's encounter caused the sickness and the air pocket formation. The sickness that Glenda felt may be attributed to the encounter as well. The delay in the onset of the sickness may be due to the fact that she did not go outside. The baby did not get sick, however.

This investigator feels that these witnesses are telling the truth. Glenda did not want to talk about her experience at first because of the sighting that her husband had may have been interpreted as the cause of her sighting. She could not resist telling her story, though. She was positve in her description.

The occurance of the second sighting on February 14 and the possibility that Jerry's illness (he has never been sick enough to be hospitalized before) was caused by the sighting, it is my opinion that this sighting *INDICATES THERE* is still a threat to the Dayton community. The public health department and state athorities should be notified.

Follow up.
Obtain the medical records from Jerry's VISIT to the hospital.
Perform further investigation in the Crosby area.
Notify athorities
Advertize for additional witnesses on radio and TV if possible.

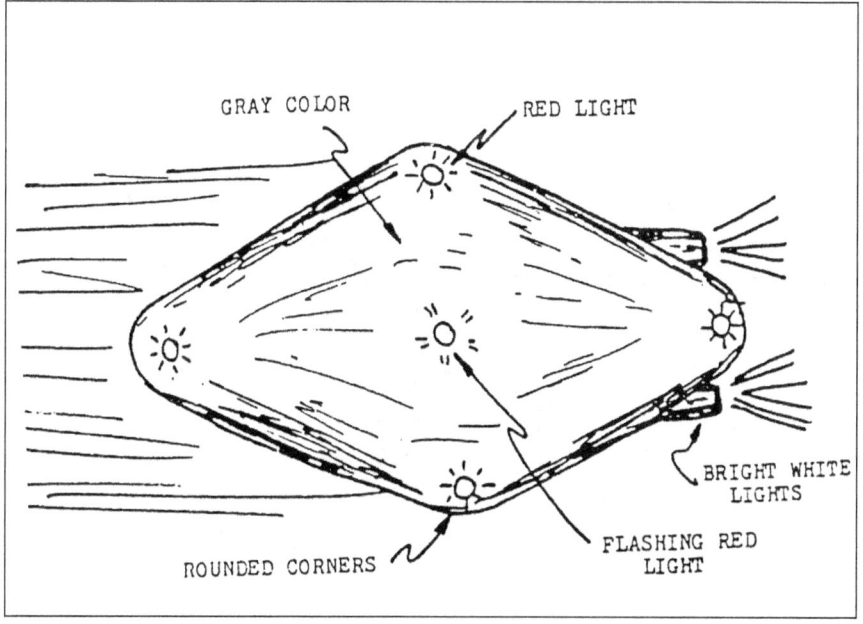

TELECON/CONFERENCE/TRIP REPORT
(CROSS OUT NON-APPLICABLE TITLE)   PAGE 1 OF 1

VISIT
Vehicle Internal Systems Investigative Team

SUBJECT/PURPOSE: CASH/LANDRUM CASE

DATE OF TELECON/TR/CONF.: 17 May 1981

DISCUSSION/COMMENTS (INFO. OBTAINED, CONCLUSIONS)

1. Vickie Landrum provided photographs of one of the CH-47 helicopters and the name of the pilot who claimed to have flown after the UFO near Huffman, Tx on 29 December 1980.

2. I met with a representative of the U.S. Army and Texas National Guard, and obtained the following information. The vehicle belonged to the 136th Transportation Company. The Commanding officer is Major Del Livingston. The unit has 16 CH-47's at Grand Prairie, Texas (Near Dallas) at the Dallas Naval Air Station. They have an additional eight CH-47's stationed at Ellington AFB, near Houston, Tx.

ACTION REQUIRED

PREPARED BY: [signature]   **DATE:** 5/17/81

P.O. Box 877 • Friendswood, Texas • 77546

TELECON/~~CONFERENCE/TRIP REPORT~~
(CROSS OUT NON-APPLICABLE TITLE)

PAGE   OF

VISIT
Vehicle Internal Systems Investigative Team

SUBJECT/PURPOSE: CASH/LANDRUM CASE

DATE OF TELECON/TR/CONF.: 3/20/81

Jerry MacDonald called J. Schuessler

DISCUSSION/COMMENTS (INFO. OBTAINED, CONCLUSIONS)

1. Metro News Service carried a plea for witnesses to come forth. Jerry heard the plea on KIKK radio.
2. Between 8 & 9 pm on 29 Dec 1980 Jerry was in his back yard working on a sewer problem. A large object came slowly across the area. It was bright, red in center & fire in back. It crept along, hovered and rumbled. It was frightening. He told only his family & close friends. It was about the altitude of twice the height of the football field light (< 200 ft. altitude). Jerry lives at 203 Hill St in Dayton, TX. telephone (713) 258-5956

ACTION REQUIRED
Visit Mr MacDonald
He lives in a trailer house with a for sale sign. When arriving in Dayton turn right at first red light and left at stop sign.

PREPARED BY: [signature]
DATE: 3/20/87

P.O. Box 877 • Friendswood, Texas • 77546

TELECON/~~CONFERENCE/TRIP~~ REPORT
(CROSS OUT NON-APPLICABLE TITLE)

PAGE 1 OF 1

VISIT
Vehicle Internal Systems Investigative Team

Name not to be used.

SUBJECT/PURPOSE: UFO Report - Conroe, Tx June - July 1980

DATE OF TELECON/TR/CONF.: 3/24/81

Bob ▓▓▓ called John Schuessler.

DISCUSSION/COMMENTS (INFO. OBTAINED, CONCLUSIONS) Phone ▓▓▓

1. Mr. ▓▓▓ called as a result of our request for eyewitnesses on KIKK radio.

2. Mr. ▓▓▓ was a security guard at the Girl Scout camp at Conroe, Tx. He was stopped by the highway entrance to the camp at 2:30 am. Suddenly, the whole sky lit up in blue light. He got out of the car, heard a strange noise, but saw nothing.

3. About 3:30 am he was standing under a tree at the camp talking to a guard from the next camp, near Pine Tree Lodge, when a big object hit & bent the top of the tree. It made a "Swoosh" noise.

ACTION REQUIRED:

PREPARED BY: John F. Schuessler   DATE: 3/24/81

P.O. Box 877 • Friendswood, Texas • 77546

TELECON/~~CONFERENCE~~/~~TRIP~~ REPORT
(CROSS OUT NON-APPLICABLE TITLE)

PAGE 1 OF 1

Vehicle Internal Systems Investigative Team

SUBJECT/PURPOSE: REPORT OF UFO NEAR CLEVELAND, TX

DATE OF TELECON /TR/ CONF: 27 APRIL 81

Darlene Vanghel Called J. Schuessler

DISCUSSION/COMMENTS (INFO. OBTAINED, CONCLUSIONS)

1. Ms. Vanghel is an ex-nurse at Methodist Hospital. Her friend Paul Hadley is the Head of Maintenance for the 3:30-11:30 shift at the complex. Mr. Hadley saw the UFO.

2. He was returning home to Cleveland from Houston after work (around midnight) in late Dec. 1980. Just past the bridge he came across a UFO at treetop level - it was the length of two football fields. He got out of the car and observed it for 20-30 minutes. Inside he could see three humanoids with "Roman noses", wearing silver suits. Their helmets seemed to be separate from the suits. They were all looking out windows in different directions. Each must have been 6-7 ft. tall. The light inside reflected off the instruments. The object made no sound. The object went up & away rapidly. The creek recked with a horrible stench after the object left.

ACTION REQUIRED

J. Schuessler to contact Paul Hadley and get his story and the date of the incident.

PREPARED BY: [signature]

DATE: 4/27/81

P.O. Box 877 • Friendswood, Texas • 77546

ON/CONFERENCE/TRIP REPORT
(CROSS OUT NON-APPLICABLE TITLE)

PAGE 1 OF 1

Charles R. (Russ) Meyer
Regional Inspector
Radiation Control Branch
Texas Department of Health
Division of Occupational Health & Radiation Control

3901 Westheimer, Suite 301
Houston, Texas 77027
7600
961-ZZZZ
Area Code 713

Project VISIT — Vehicle Internal Systems Investigative Team

SUBJECT/PURPOSE: CASH/LANDRUM CASE

J. Schuessler met with Russ Meyer, G. Freeland, and M. Vredenburg of the Radiation Control Branch of Texas Dept. of Health per V. Landrum request

DATE OF CONF.: 9/10/81

DISCUSSION/COMMENTS (INFO. OBTAINED, CONCLUSIONS):

1. State Representative Browder requested the above noted group to look into the case, as the result of Betty Cash & Vickie Landrum visiting his office in August. Russ Meyer is a Regional Inspector and acted as spokesman for the group.

2. I gave them an overview of the incident and the injuries. They were quite interested, but said they couldn't prove whether or not radiation had been present. This type of case is not their normal job. They would like to have a doctor tell them whether or not his opinion is radiation in this case. They would report that opinion back to the Representative in Austin. It would then be up to him to pursue the source, type, responsibility, etc. They could send a report to the State Medical Advisory Board in Austin but it only meets once a year. They felt witnesses would live, since they are now alive.

ACTION REQUIRED:

They will look at Betty's hospital records if she wants to give permission.
They suggested going to Dr. Vince Collins, Rosewood Hospital, for his opinion on the case.

PREPARED BY: J. Schuessler   **DATE:** 9/10/81

P.O. Box 877 • Friendswood, Texas • 77546

ABOUT THE AUTHOR

JOHN SCHUESSLER has been involved in the United States manned space program since 1962 and a UFO researcher since 1965. He is a founding member of the Mutual UFO Network, Inc., is presently the Deputy Director for Administration, a consultant in Astronautics, and a member of the Board of Directors. As a staff member, he has written numerous articles for *Skylook* and the *MUFON UFO Journal* and featured at MUFON symposia eight times. He administers the MUFON Medical Committee, composed of consultants with medical degrees.

He is currently a member of the UFO Research Coalition Board of Directors. He was a founding member and past President of the UFO Study Group of Greater St. Louis, Inc. He was also a founding member of the Houston-based Vehicle Internal Systems Investigative Team (VISIT). He is a member of the Center for UFO Studies (CUFOS) and the Houston UFO Network (HUFON).

John is employed in the aerospace industry and has been involved in the engineering side of most of the U.S. manned space programs, from Project Mercury to the International Space Station. His primary field of interest is the technology of the future. He holds a Master of Science degree in Studies of the Future, Technology Forecasting from Houston Clear Lake.

John has demonstrated a long-standing interest in advanced propulsion concepts, as indicated by the data in many UFO reports. He has approached his research in this area by examining the effects of UFO close encounters on human systems. In 1996, he cataloged his research in a book entitled *UFO-Related Human Physiological Effects*. He has appeared on numerous radio and television shows such as *That's Incredible, Good Morning America, Unsolved Mysteries, In Search Of,* Sightings, and *Strange Universe*.